D0221159

TELECOMMUNICATIONS AMERICA

New Titles from QUORUM BOOKS

The Export-Import Bank at Work: Promotional Financing in the Public Sector
Jordan Jay Hillman

Supply-Side Economics in the 1980s: Conference Proceedings
Federal Reserve Bank of Atlanta and Emory University Law and Economics Center, Sponsors

Deregulation and Environmental Quality: The Use of Tax Policy to Control Pollution in North America and Western Europe
Craig E. Reese

Danger: Marketing Researcher at Work
Terry Haller

OPEC, the Petroleum Industry, and United States Energy Policy
Arabinda Ghosh

Corporate Internal Affairs: A Corporate and Securities Law Perspective
Marc I. Steinberg

International Pharmaceutical Marketing
Suresh B. Pradhan

Social Costs in Modern Society: A Qualitative and Quantitative Assessment
John E. Ullmann, EDITOR

Animal Law
David S. Favre and Murray Loring

Competing for Capital in the '80s: An Investor Relations Approach
Bruce W. Marcus

The International Law of Pollution: Protecting the Global Environment in a World of Sovereign States
Allen L. Springer

Statistical Concepts for Attorneys: A Reference Guide
Wayne C. Curtis

Handbook of Record Storage and Space Management
C. Peter Waegemann

Industrial Bonds and the Rating Process
Ahmed Belkaoui

Of Foxes and Hen Houses: Licensing and the Health Professions
Stanley J. Gross

TELECOMMUNICATIONS AMERICA

Markets Without Boundaries

MANLEY RUTHERFORD IRWIN

KF
2765
I78
1984

Q

QUORUM BOOKS
WESTPORT, CONNECTICUT
LONDON, ENGLAND

EERL Withdrawn
RASMUSON LIBRARY
Surplus/Duplicate
UNIVERSITY OF ALASKA-FAIRBANKS

Library of Congress Cataloging in Publication Data

Irwin, Manley Rutherford
  Telecommunications America.

  Includes Bibliographical references and index.
    1. Telecommunication—Law and legislation—United
States. 2. American Telephone and Telegraph Company.
3. Antitrust law—United States. 4. Consent decrees—
United States. 5. Telecommunication policy—United
States. 6. Telecommunication—United States. I. Title.
KF2765.I78   1984   343.73′0994   83-9448
ISBN 0-89930-029-4 (lib. bdg.)   347.303994

Copyright © 1984 by Manley Rutherford Irwin

All rights reserved. No portion of this book may be
reproduced, by any process or technique, without the
express written consent of the publisher.

Library of Congress Catalog Card Number: 83-9448
ISBN: 0-89930-029-4

First published in 1984 by Quorum Books

Greenwood Press
A division of Congressional Information Service, Inc.
88 Post Road West
Westport, Connecticut 06881

Printed in the United States of America

10 9 8 7 6 5 4 3 2 1

To
Doris, Trisha, Douglas

CONTENTS

TABLES AND FIGURES

TABLES

FIGURES

PREFACE

This study is derived from a recent course introduced at the Whittemore School of Business and Economics—"Technology, Information and Public Policy." Dwight Ladd, Dean of the Whittemore School, provided seed money for the course and subsequently encouraged the effort that went into this present study.

Discussions with John Ela, a recent graduate of our Masters' program, formed and crystallized many of the ideas and concepts included in this study. I am indebted for his help and assistance.

Douglas A. Irwin served as counselor, advisor, researcher, and editor of the entire study. This book would not be the same without his inspiration and help.

Finally, Madeline Piper has typed the book's manuscript from beginning to end many times. My gratitude for her skill and patience.

ABBREVIATIONS

ACS   Advanced Communication Systems
AT&T   American Telephone and Telegraph
ATM   automatic teller machine
BTL   Bell Telephone Laboratory
BSPPD   Bell Systems Purchase Product Division
CCSA   common control switching arrangement
CSO   central staff organization
CWA   Communications Workers of America
DDP   distributed data processing
DTS   digital termination system
ECOM   electronic computer originated mail
EFTS   electronic funds transfer
ENFIA   Exchange Netword Facilities for Interstate Access
FCC   Federal Communications Commission
FERC   Federal Energy Regulatory Commission
FX   foreign exchange
GTE   General Telephone and Electronics
IBEW   International Brotherhood of Electrical Workers
IC   integrated circuit
ICA   International Communications Association
ICC   Interstate Commerce Commission
IFOC   International Fiber Optics and Communications
ITT   International Telephone and Telegraph
LATA   local access and transport areas
MCI   Microwave Communication Incorporated
MTS   message toll telephone service
NARUC   National Association of Regulatory Utility Commission
NPR   National Public Radio
OTA   Office of Technology Assessment

OTP	Office of Telecommunication Planning
PBX	private branch exchange
PUC	Public Utility Commission
R&D	research and development
RCA	Radio Corporation of America
SMATV	satellite master antenna television
TNO	The Netherlands Organization for Applied Scientific Research
TNS	Transaction Network Service
USITA	United States Independent Telephone Association
USPS	United States Postal Service
VLSI	very large scale integration
WUTC	Washington Utilities and Transportation Commission

1

INTRODUCTION

Within the past two years, the United States has undergone more changes in its telecommunications industry than in the previous seven decades. American Telephone and Telegraph (AT&T), the dominant domestic carrier, has agreed to sell over $80 billion worth of assets, to part with one-third of its revenues, and to separate itself from 800,000 employees in the largest single divestiture in U.S. corporate history.[1] Clearly, U.S. telecommunications, its policies, practices, and direction, will never be quite the same again.

This study explores the implications of the AT&T-Department of Justice Consent Decree of 1982. In so doing, it

- summarizes briefly the world of pre-1982 regulation
- identifies the forces that eroded U.S. telecommunication policies and practices
- documents Bell's response to this change and environment
- outlines the content of the 1982 Bell Consent Decree
- speculates as to the future direction of information products and services in the United States
- searches out the new mandate of regulation at both the state and federal level
- juxtaposes future trends of the economy and the drift of regulation.

Today there is a growing mismatch in the United States between the reality of an information infrastructure emerging in our economy and the thrust of telecommunication regulation. Technology is dissolving industry boundary lines, assaulting geographic demarcations, eroding sector distinctions, and softening global artifacts—all of which heralds the prospect of intensified domestic and global competition. In an environment of rapid change, the future holds unlimited opportunities and yet embodies unprecedented risks. An emerging environment may well belong to the entrepreneur and the venture capitalist.

While all this is happening, federal and state regulation enlarge and broaden their jurisdictional reach. It is one thing to assert that telecommunication regulation is out of control; it is another to observe that utility oversight is an institutional anachronism. It is the thesis of this book that regulation as currently structured threatens to dissipate the energy and creativity of individuals seeking to confer more, rather than fewer, options to the economy as a whole.

MONOPOLY AND BOUNDARIES

For decades, the premise of telephone monopoly and geographic exclusivity appeared not only workable but also superior to any alternative form of industrial organization. American Telephone and Telegraph and the Bell System epitomized that structure. As a holding company, AT&T controlled twenty-two Bell operating companies scattered throughout the country; owned its Long Lines Division that provided toll telephone service linking the various Bell and non-Bell operating companies; owned Western Electric, a supplier of equipment and telephone apparatus; and operated Bell Telephone Laboratory, the research arm of the entire system.

The Bell System's policies dominated the U.S. scene to the extent that AT&T accounted for about 80 percent of all local exchange calls, upward of 90 percent of all long distance calls, the bulk of telephone manufacturing equipment, 20 percent of all corporate equity, and some 5 percent of all corporate debt. The policies regarding terminals, services, and equipment became the norm for the industry and set the pattern for non-Bell Telephone companies throughout the United States. The telephone instrument was leased rather than sold to the telephone subscriber—a policy derived, according to AT&T, from a mandate for end-to-end communication service. The telephone company's service responsibility, it said, rested on the ability to own, repair, and upgrade the telephone instrument as part of a delicately interconnected service linking millions of subscribers.

Exclusivity of local exchange service, indeed long distance service, fell within the constraint and content of a natural monopoly. Competition resulted in the duplication of exchange facilities and was held to be impractical given the high start-up costs and the demand for universal telephone service.

Vertical integration, the ownership of utility and manufacturing, was concentrated within AT&T's holding company format—specifically, AT&T's Western Electric affiliate. The Bell operating companies bought most of their equipment from Western or via Western.

The Bell System's practices mirrored the industry structure of AT&T—namely, Bell Laboratory designed hardware and equipment, Western Electric manufactured the products, and Bell operating companies bought hardware as an investment associated with the delivery of service to their subscribers. Acting as master coordinator, AT&T balanced the needs and the working relationship of the constituent parts that ultimately became known as the Bell System. Although non-Bell or independent firms and manufacturers were important players on the American scene, they were, in a relative sense, overshadowed by the size and organization of the Bell System.

A second premise of the U.S. telecommunications system held that industry's demarcations of products, services, and geography be clear-cut and readily identifiable. To the extent that the firm possessed a product, service, or geographic exclusivity, it was the epitome of a natural monopoly and thus invited some form of accountability. As a result, the public utility commission became institutionalized as an instrument of social control. Public utility regulation commenced with the railroad industry and was extended to telephone and telegraph carriers at the federal level by 1910. The Communication Act of 1934, in creating the Federal Communications Commission (FCC), merely formalized an institution whose roots had evolved at the state level by the first decade of this century.

Given our federal system, regulation was divided between state agencies and the FCC. Within that division, each agency pursued "rate base economics" in determining the costing, pricing, and profits of their respective carriers. Under this system, a regulated utility was entitled to recover operating expenses as well as permitted to generate a return on net capital investment. This return or profit, equated to the cost of capital, was inserted as a markup over capital investment. Cost plus thus became a determinative mechanism for regulated carriers or public utilities.

Regulation attempted to balance the carrier's financial need against the subscribing public's requirements for universal telephone service. That balance constituted objective and subjective decisions resolved through a complex legal system of due process and judicial review.

Over time, technology began to smile at long distance costs and to frown on local exchange costs. As productivity translated into lower costs at the toll end of telephone service, state commissions argued that revenues from long distance be transferred to support rates at the local exchange level. Thus began a series of internal subsidies of incredible complexity (toll versus local, business versus residential, urban versus rural) that permitted the allocation of joint plant, sanctioned and approved by both state and the FCC.

These subsidies flowed within markets whose boundaries and demarcations were clear and accepted by all. No one confused the differences between telephony, mail, telegraphy, office, the typewriter, radio, television set, or whatever. Product, service, and geographic boundary dimensions were fixed and stable—so stable, in fact, that in 1956 AT&T agreed to confine its service and manufacturing to regulated communication common-carrier services only.[2]

This agreement arose from an antitrust settlement with the Department of Justice that had sought to spin off Western Electric as an overextension of Bell's monopoly in telephone manufacturing. Through its retention of vertical integration, Bell agreed to confine its services and diversification to regulated markets, a decision that at the time was perceived as an exercise in management perspicacity. The decree saved the integrated nature of AT&T and acknowledged that the constituent parts of that system were subject to some type of regulatory protection.

DISCONTINUITY

Alfred Marshall, the neoclassical economist, once observed that nature makes no leap but rather engages in small steps. His observation appears no less compelling today. Changes in the state of the telecommunications art occur at the margin, gradually eroding the presumption of monopoly and the premise of immutable industry boundary lines.

Interestingly enough, the legitimacy of total monopoly as well as that of firm practices first surfaced at the federal policy front. In the 1960s, a presidential task force on communications, an FCC investigation of the use of computers over telephone lines, and a congressional debate over the nation's ownership of satellites broached questions that had been regarded as unthinkable in decades past.[3] In the middle 1960s, a presidential task force could see no major problem in permitting subscribers to tie equipment to telephone lines. The task force even encouraged the entry of specialized carriers into long distance leased facilities competitive with AT&T. Under congressional pressure, COMSAT, a space consortium, was legislatively required to purchase its equipment under competitive bidding supervised by the FCC. The FCC in 1968 and again in 1969 encouraged subscriber ownership of equipment and approved the licensing of a specialized carrier.

These decisions by Congress, the executive branch, and the regulatory agencies precipitated a conflict between the philosophy of state regulation and the direction of the federal regulatory action. The injection of competition into formerly monopolized markets signified a break with the concept of rate base, end-to-end service, monopoly, and total industry control. Furthermore, competition led to an erosion of the rate base and of long distance revenues which ultimately could dry up contributions to the local exchange rate. In the absence of the toll subsidy for local exchange, rates would inevitably rise. Not surprisingly, state regulators stood united in opposing any dilution of toll subsidy at the exchange level.

A second development also fell within Marshall's dictum of incremental change. What appeared to be immutable industry boundary lines between and among firms was now beginning to soften under the pressure of microelectronics, computers, satellites, office electronics, banking, automation, fiber optics, broadcast satellites, telecommunications, and cable. In the 1970s, an era of rampant inflation, the ravages of technological deflation nibbled at traditional demarcations and industry separations.

Industries that were once unrelated began to vector on a collision course; past characteristics distinguishing the laboratory, office, factory, home, store, and bank were no longer as clear-cut as in the past. More significantly, technology opened up what appeared to be the unlimited potential of an information economy that linked many industrial sectors within the United States. The revenue prospects of information products, networks, and services were not only staggering but also patently inviting to a host of industries including telephony.

RESPONSE OF THE BELL SYSTEM

The challenge to market entry and the erosion of market boundary lines began to affect the U.S. telecommunication industry, and AT&T in particular. AT&T saw its markets and services experiencing outside technological and market entry, thus eroding its longstanding monopoly status. Bell opposed terminal ownership by subscribers, fought equipment certification, dropped its equipment rates, bundled telephone equipment and lines, imposed interface devices upon interconnect or non-Western Electric suppliers, opposed the premise of specialized carriers, refused to lease local service loops, and then agreed to sell loops, cut prices on private line services by 80 percent, and limited the extent of leasing lines to its competitors on the premise that their service content exceeded regulatory sanction.

AT&T continued to salute the virtues of vertical integration, buying the bulk of its equipment requirements from Western Electric. AT&T's buying practices continued despite new suppliers, new equipment, lower costs, alternative features that served as a challenge to the products, and the strategy of vertical integration.

Bell's defense of terminals, lines, and integration eventually produced a rash of antitrust complaints. Competitors alleged that AT&T was engaging in monopoly and predatory pricing in violation of the U.S. antitrust statutes, and they sought redress through the courts, if not through the regulatory agencies. In some cases, AT&T won; in other cases, Bell's opponents emerged victorious; and in still others, the suits were settled out of court.

In 1974, however, the Department of Justice filed an antitrust complaint alleging that AT&T had monopolized the terminal market, the private line market, and the equipment manufacturing market. The Department of Justice vowed that nothing less than massive corporate surgery applicable to the entire System could ever hope to return the industry to one of equitable access and fair competition.

As if market entry within one's own turf were insufficient, AT&T began to realize that technological possibilities were opening up and beckoning new markets, services, and products. In the 1970s, for example, AT&T introduced a computer terminal as an upgrade of its hard copy printer that it had manufactured through Western Electric for over thirty years. Such an upgrade was inevitable, and under its Consent Decree mandate, AT&T filed appropriate tariffs before the FCC and state regulatory agencies.[4]

Then the expected unfolded. The computer industry opposed AT&T's filing on the grounds that a computer terminal was in violation of Bell's Consent Decree and that AT&T could not justify any competitive or unregulated markets unless sanctioned and approved by the Department of Justice. The Department of Justice expressed absolutely no interest in lifting the 1956 Consent Decree.

The case of the computer terminal known as the Teletype Model 40 arose before the FCC. In a close vote, the FCC did agree to accept the Teletype Model

40 as an appropriate regulatory activity. The decision was critical. Without regulatory sanction, however, both AT&T and Western Electric would have been inhibited in their ability to diversify and move into new information products.

As AT&T included more microchips in a range of telephone equipment, hardware, and products—adding computer, data processing, and storage capability whether in packet switching terminals, PBXs, or data modems—the question of whether Bell violated the Consent Decree and the boundaries of common-carrier telephony, and whether AT&T was diversifying, however silently, into data processing or computer activity became more striking and controversial. Technology continued to soften market boundary lines; the 1956 Consent Decree slowly began to move from the status of corporate genius to that of a management nightmare. AT&T tried to finesse its diversification by applying the term "communication processing" to new equipment and services introduced before the FCC or the state regulatory agencies. Nevertheless, the computer industry remained doggedly determined to oppose any AT&T diversification on the grounds that data processing resided outside the agreement reached with the Department of Justice in 1956.

Any new service sanctioned by the state or FCC also found itself spilling into regulatory controversy and ultimately toward judicial review. If due process existed as a carrier asset to delay potential competitors under the old system, due process also began to shift into a handicap impeding the ability of the telephone company to meet market changes with new products, services, and ideas. In sum, the antitrust suit and the constraints placed upon the diversification by the 1956 Consent Decree set in motion the Consent Decree of 1982.

THE 1982 CONSENT JUDGMENT

The agreement between the Department of Justice and AT&T has recently been approved by the presiding district court judge. By 1984 AT&T will divest itself of twenty-two operating companies, companies that will spin off into seven separate regional telephone associations. AT&T's Long Lines Division will supply long distance service not only between states but within state boundary jurisdictions as well. In addition, AT&T will retain Western Electric and Bell Telephone Laboratories.

In return for the divestiture of twenty-two operating companies, the Consent Decree of 1956 will be lifted and AT&T will be permitted (and indeed has established) a separate subsidiary to engage in nonregulated services and equipment—that is, to diversify into products, services, and networks outside the reach of traditional common-carrier regulation. Under pressure from the newspaper industry, the judge has barred AT&T from engaging in electronic publishing services for a period of seven years.

In addition to retaining local exchange service, the Bell operating companies now divested will possess an enlarged service area and will retain the yellow pages as well as the ability to buy and sell customer premise telephone equipment.

Interconnect suppliers of equipment, that is, those who sell terminals directly to telephone customers, will enjoy market access through the Bell operating companies. Specialized carriers will presumably be given access to the local distribution facilities of Bell operating companies on the same basis as AT&T. And some time in the future, non-Western Electric manufacturers should find the market for Bell operating companies an attractive one in terms of equipment, products, hardware, and apparatus.

All of this marks a striking transition of an industry from telephony to information services and equipment. In some respects, this transition acknowledges that a new environment imposes a new reality upon the firm and its constituent industries. While it is still early to define the precise content of that environment, a rough configuration can be detected.

THE FUTURE INFORMATION ENVIRONMENT

Perhaps it is easier to address the outline of this new environment through a series of questions. First, will the base, range, and diversity of technology taper off in the foreseeable future? So far, the answer apparently is no. One can discern in the microelectronic revolution new generations of satellites, germanium, arsenide, lithography, computer architecture, fiber optics, optical switching, hardware terminals, office automation, and so on. The range of alternatives in technology appears to be increasing, and the possibility of coupling, linking, and interacting these choices is growing rather than contracting, or, as some observers put it, we are just now experiencing a technological explosion.[5]

Second, is this technological genie out of the bottle both internationally and within the domestic arena of the United States? Thus far, the answer appears to be yes—whether in computers, software, satellites, or aerospace, technology is rampant in the Far East, North America, and Europe. The dissemination of research, expertise, knowledge, and talent in microelectronics is no longer geographically reserved for one or two companies, much less for one or two nations. One can discern an international division of labor in information products, services, and networks. No one firm—indeed, no one country—possesses research and development exclusivity.[6]

Third, if traditional market boundary lines are decaying—at least in the United States—is there any indication that these boundary lines will be hard and crystallized in the future? Perhaps, but the characteristics unique to telephone, banking, finance, telecommunications, computers, aerospace, retail, and broadcasting are dissolving quickly. Driven by technology, networks, products, and services, the post office, telephone company, broadcaster, cable TV, and computer firm are beginning to trend toward the same markets, customers, and revenues.

Fourth, if falling market boundary lines constitute a continuing trend, will our environment experience more rather than less competition? The odds are that it will experience more rivalry. As adjacent industry boundary lines dissolve, firms whose relationship in the past was distant or even unrelated now confront each other as direct competitors and rivals. In the United States, American Satellite, for example, competes with Satellite Business Systems (IBM): the airlines,

Western Union, Federal Express, Holiday Inn, the broadcast network, specialized carriers, and cable TV companies. Competition shows every sign of intensifying rather than atrophying in the remaining years of the 1980s.

Fifth, will intensified competition spur economic performance, whether price or nonprice? There is mounting evidence that innovation will accelerate in the foreseeable future. The economic life of equipment and hardware, whether in ICs, satellites, PBXs, exchanges and a variety of office equipment, consumer electronics, will contract and shorten.[7] The rate of obsolescence—the inverse of innovation—is obviously picking up as well. Hence, profit volatility will migrate to more industries as part of a dynamic environment of intensified price competition.

Yet, price competition represents one facet of this new environment. Commodity-like products will shift to differentiated products. Segmentation and specialization will be critical factors, and the transition of productivity to price will border on the furious, injecting more fluidity and change in an environment already beset by rampant change. Spawned by cost reductions and price elasticity, global competition injects still another element of dynamism into this environmental reality.

Sixth, if market innovation poses as one key to corporate decision-making, what corporate structure is most apt to thrive in an environment of risk, uncertainty, and change? The returns are obviously not in, but young companies, small firms, the entrepreneurs, and venture capital enjoy a strategic role in an environment of change, adjustment, and agile response.

Venture capital possesses at least one virtue—the ability to respond, innovate, and introduce products from the laboratory into the marketplace. Certainly, that has been the experience in voice mail, home computers, super minicomputers, on-line word processing, digital PBXs, computer-aided design, local area networks, fail-safe computers, computer peripheral systems, digital satellite relay, diagnostic data modems, smart multiplexors, and packet switching. Alternatively, one might speculate that any organization with a rigid hierarchy may no longer possess the sole corporate model for successful performance in a decade of accelerated change.

If the elements of this new market are valid, they can be summarized as follows:

- There appears to be no foreseeable letup in the microelectronic revolution.
- The technological revolution is now global in dimension, no longer confined to one firm or one nation.
- Industrial products and service boundary lines are softening and dissolving.
- Geographic markets are less insulated than in the past.
- Intensified competition within and among industries can be anticipated to be a new market reality.
- Price and nonprice decisions will be under pressure to accelerate.

●The entrepreneurial organization possesses advantages in an environment of innovation and change.

THE FUTURE OF REGULATION

The Federal Communications Commission

In the post-AT&T/Justice settlement era, both the FCC and state regulations will experience unprecedented adjustments inherent to an environment of economic turbulence. Under the terms of the settlement, the FCC has been assigned the mandate to monitor the division between AT&T's regulated and unregulated activities. If that demarcation remains static and clear-cut, then the FCC's oversight may be reduced to a mechanical exercise. But if the boundaries are shifting, what may appear to be susceptible to rate base economics today may well lend itself to deregulation tomorrow. Will the commission—indeed, any institution—consign itself to oblivion? The answer is probably not.

A more likely prospect is that information, whatever its content—electronic funds transfer, remote bibliographic services, on-line word processing, interactive videotext, office automation, or home information services—may experience increased regulatory overview. The commission's definition of the "public interest" is so broad as to accord the agency almost unlimited jurisdiction. Witness the FCC's quarrel with the U.S. Post Office's use of satellites for facsimile transmission.[8]

Another question is, how will federal regulation monitor a bifurcated firm straddling regulated and nonregulated activities? Will the imposition of accounting and financial standards lend clarity to the joint assets of Western Electric and Bell Laboratories, to say nothing of AT&T? Will the hiring of additional FCC staff personnel put the problem of joint cost behind us?[9] Will not the fear of "cross-subsidization" at the manufacturing and research level provide a new rationale for more rather than less regulatory oversight, not unlike the Canadian experience?

The answers to these questions remain pending. But if definitional clarity has proved elusive at the services end of the business for twenty years, despite the FCC's attempt to promulgate over fifteen new boundary demarcations, how can any regulatory agency—no matter how well intentioned—fine tune vertical demarcations in manufacturing and research and development (R&D)? (Western Electric, for example, manufactures over 200,000 different products.) Is it possible that regulation prefers to be consumed by due process rather than measured by economic results? If the former, then the next ten years will merely replicate a dialogue of semantics, definitions, guidelines, standards, and commission exhortation.

How will the FCC monitor the conduct of a remaining vertically integrated AT&T Long Lines/Western Electric organization? The Department of Justice suggests that a 5 percent penetration by specialized carriers in the long-haul market will spring open a foreclosed AT&T/Western Electric equipment market sometime in the future. Is that premise valid? Perhaps. But the commission's stance in fiber optic communications may be prophetic. The FCC permitted AT&T to allocate its initial fiber optic requirements to Western Electric exclusively; then requested Bell to entertain competitive bidding to outside vendors; then witnessed a Japanese firm submit the low bid; then sanctioned Bell's withdrawal of the award, reassigning optics to its own affiliate on the basis of "national defense"; and then permitted AT&T to stipulate that non-Western Electric fiber purchased by AT&T must be manufactured in the continental United States.[10] When someone suggested that vertical integration might foreclose the next generation of transmission facilities, the commission noted that AT&T's subcontracting policy would "provide a more competitive telecommunications equipment manufacturing industry."[11]

Finally, can a regulatory agency promulgate meaningful policies in an environment beset by heavy doses of domestic and global rivalry? Does not the premise of the public utility process stand at odds with the reality of an information economy?

State Regulation

The burden placed upon state regulatory commissions in the United States is no less imposing. First, how will the state adjust to an era of lost revenue subsidies? Will the local access charge be sufficient? Will not state commissions be under enormous pressure to lift those charges in order to prevent rapid rate increases at the exchange level? Is it possible that high access charges will drive some firms to seek alternatives to bringing their services to the home, office, or store via telephone facilities? Certainly the options, both existing and potential, are expanding—satellites, fiber optics, direct broadcast satellites, digital broadcasting, cellular radio, coaxial cable, and so on—and the range of potential suppliers is also increasing—specialized carriers, other telephone companies, cable TV, satellite carriers, and value-added carriers. And is it not conceivable that AT&T itself, coerced by market forces, may find itself employing technology to circumvent the plant investment of its former telephone operating companies? Will not that endeavor place AT&T in opposition to fifty state public utility commissions? Perhaps state regulatory agencies will be placed in the position of banning new technology outright in the name of preserving subsidies derived from the local access charge.

All of this is speculative. Yet the California Public Utility Commission asked the district court to foreclose AT&T from bypassing local operating facilities. The state commission cited the following bypass possibilities:

- large corporate headquarters in a single building or a single office complex

- new industrial or commercial tracts or large office buildings owning their own communications networks within the tract or building and furnishing services to all occupants

- large scale operations having a high volume of traffic to coordinate with computers in other cities
- large government office concentrations
- university campuses[12]

(The commission neglected to include the use of carrier pigeons bearing microfilm copies of graphic design 30 miles between offices of the Lockheed Corporation.)[13] In any case, Judge Harold Greene of the district court rejected the state public utility agency request.[14]

Can state regulatory agencies adjust to an era of rate structure—the matching of specific price to cost—rather than the tradition of determining overall revenue requirements? That undertaking will be proven to be as formidable as it is complex. Perhaps the FCC's experience in struggling with fully allocated versus incremental cost may prove enlightening to state agencies. In any case, U.S. regulation at the state level will never be quite the same.

SUMMARY

The AT&T/Department of Justice Consent Decree marks an historic step in U.S. telecommunications. In many respects, the emergence of an information infrastructure and the legacy of public utility regulation stand in striking opposition. One is static and backward looking; the other is dynamic and forward looking. AT&T's decision to divest its operating companies in order to secure flexibility and information markets marks an attempt to grasp the opportunities of the future.

Less noticeable is the fact that AT&T distanced itself from fifty state public utility commissions, organizations closely identified with the companies' well-being for the past half-century. In one sense, state regulation both lost and retained a client. Is this loss of constituency the beginning of a long-term process?

NOTES

1. United States v. American Telephone and Telegraph, Western Electric Company, Bell Telephone Laboratories, Civil Action No. 74-1698, U.S. District Court, District of Columbia, August 11, 1982.

2. U.S. Congress, House, *Consent Decree Program of the Department of Justice*, Hearings Before the Antitrust Subcommittee (Subcommittee No. 5), Committee on the Judiciary, 85th Cong., 2d sess., Part 2, American Telephone and Telegraph Co., 1958.

3. U.S. President's Task Force on Communication Policy, *Final Report*, U.S. Government Printing Office, December 7, 1968.

4. FCC, In the Matter of AT&T's Revision to Tariff FCC Nos. 260 and 267, *Memorandum, Opinion and Order*, March 3, 1976, p. 7: "The Dataspeed 40/4 is thus an offering of data processing services."

5. Kojikobayashi, "Telecommunications and Computers: An Inevitable Marriage," *Telephony*, January 28, 1980, p. 78; H. Hindin, "Large Scale Integration Latches onto the Phone System," *Electronics*, June 5, 1980, p. 10; and M. R. Irwin, "Telecom User Pressure Is Changing Market Boundaries," *Telecommunications* (September 1982):17.

6. Adam Osborne, *Running Wild: The Next Industrial Revolution* (Berkeley, Calif.: McGraw-Hill, 1979); K. G. Corfield, "Into the World of Broadcast Systems," in J. Lighthill et al., *Communications in the 1980s and After* (London: The Royal Society, 1979), p. 121; S. T. McClellan, "Sea Changes in the Information Industry," *Datamation* (June 1982):89; and "The Japanese In-

vasion: Chips Now, Computers Next,'' *Electronic Business* (July 1981):84; See also "Tomorrow's Leaders: A Survey of Japanese Technology,'' *The Economist,* June 19, 1982, p. 1.

7. M. Knox, "PBX Life-cycle Called Key to Telecom Industry Future,'' *Management Information Systems Week,* August 19, 1981, p. 11: "The life cycle of switches will follow the computer industry much more closely, where every four years the computer industry produces a new generation that is cheaper and easier to maintain.'' See also M. R. Irwin, "Technology and Telecommunications: A Policy Perspective for the 80's,'' Working Paper No. 22, Regulation Reference, Economic Council of Canada, Ottawa, Canada, p. 42; and M. R. Irwin, "U.S. Telecommunications: Technology vs. Regulation,'' *Information Society: Changes, Chances, Challenges,* Fourteenth International TNO Conference, Rotterdam, The Netherlands, 1981, p. 47.

8. R. Wiley and D. Adams, "Should the FCC or USPS Control Electronic Mail,'' *Legal Times of Washington,* April 16, 1979, p. 12. "The FCC concluded that its own enabling statute (The Communications Act of 1934) vests in it exclusive federal jurisdiction over all forms of electronic communications, including related physical delivery functions.''

9. General Accounting Office, "Can the Federal Communications Commission Successfully Implement Its Computer II Decision,'' Washington, D.C., January 29, 1982, p. 5.

10. FCC, In the Matter of Application of American Telephone and Telegraph Company, the Bell Telephone Company of Pennsylvania, the Chesapeake and Potomac Telephone Company, the Chesapeake and Potomac Telephone Company of Maryland, the Diamond State Telephone Company, New England Telephone and Telegraph Company, New Jersey Bell Telephone and Telegraph Company, New York Telephone Company, and the Southern New England Telephone Company, For authority under Section 214 of the Communications Act of 1934, as amended, to supplement existing facilities by construction, acquisition and operation of a light guide cable between cities on a main route between Cambridge, Massachusetts and Washington, District of Columbia with extension light guide cables to various cities along this route. F 16 No. W–P–C–3071. *Memorandum Opinion and Order,* April 29, 1982.

See also M. J. Richter, "FCC Rebuffs Fujitsu; OK's AT&T Fiber Pact to WE: Will Monitor Outside Buys,'' *Electronic News,* April 3, 1982, p. 1; and E. Meadows, "Japan Runs into America Inc.,'' *Fortune,* March 22, 1982, p. 56.

11. *Telecommunications Reports* 48, No. 18 (May 3, 1982):17.

12. United States v. American Telegraph Company and Bell Telephone Laboratories FWC, Civil Action No. 74–1698, Brief of the People of the State of California and the Public Utilities Commission of the State of California on Issues Regarding the Absence of Restrictions on AT&T, June 13, 1982.

13. "Carrier Pigeons Ferrying Lockheed Microfilm,'' *New York Times,* August 19, 1982, p. 20.

14. United States v. American Telephone and Telegraph Company, *Opinion,* Judge H. Greene, August 11, 1982, p. 78.

2

HISTORY OF TELEPHONY— THE BELL SYSTEM

It is exceptional to find an industry that constitutes the history of a firm. Such is the case in telecommunications and the Bell system. The origin of telephony and the Bell System begins with the basic telephone patent license.

TECHNOLOGY

On February 14, 1876, two separate individuals applied for patents on the basic telephone instrument: Alexander Graham Bell, who submitted his patent for the electric telegraph, and Elishia Gray, who was a candidate for transmitting local sounds over wires. Therein was born a controversy as to who was the first applicant, who was the original inventor, and who conceived of telephony as it has become known today.

Scribbled notes on Bell's application, later testimony by Patent Examiner Zenas F. Wilbur informing Bell's attorney about Gray's application, Bell's later examination of the Gray application, alterations in the Bell filing—all provided the ingredients for a longstanding controversy.[1] Gray was later to assert that a "golden clasp" had sealed the lips of those who knew the real story of the telephone patent.[2] By a vote of four to three, the U.S. Supreme Court upheld Bell's basic patent claim in one of the most far-reaching property awards in U.S. history.[3]

After the award, the Bell Patent Association approached the most prestigious U.S. corporation in communications, Western Union Telegraph Company, and offered this $40 million company the basic patent for $100,000. Western Union rejected the patent because, as one Western Union spokesman was to recall later, "We were telegraph men."[4] By force of circumstance, the Bell Patent Association found itself in the telephone business.

Bell sold franchises and instruments and carved out telephone districts. By 1878, crude switchboards or exchange services were introduced. Yet, Bell did not stand alone. Western Union, realizing its blunders, secured its own patent at a cost of $2 million, paying $100,000 to Thomas Edison alone. Through Edison the telegraph company had access to the most renowned research laboratory of that time, Edison's laboratory at Newark, New Jersey. Edison's microphone gave the telegraph company an early lead in the telephone transmitter. Bell acquired the Blake transmitter, however, which equaled the quality of Edison's device.

Competition for markets erupted early. Western Union's subsidiary, the American Speaking Telephone Company, served 50,000 customers in fifty-five cities, including New York and Boston. Western Electric, Western Union's equipment supplier, not only held important patents to the cord switcher and telephone instrument, but also was perhaps the most outstanding manufacturer of telephone and telegraph apparatus in the United States.

Western Union enjoyed strategic office locations in railroad stations, business establishments, hotels, and stock exchanges. When Bell licensee companies attempted to cross the rights of way or user facilities of Western Union, such access was denied. In a word, the small telephone company encountered formidable competition across the board—in research, manufacturing, and service. The Bell companies found themselves arrayed against the Edison laboratory, Western Union's rights of way, and the telegraph company's access to international markets, to say nothing of Western Union's longstanding reputation in the business community. The Bell association possessed one asset, however: the basic patent for the telephone.

Negotiations between Western Union and Bell began around 1878. Theodore Vail, recruited from the U.S. Post Office, joined American Bell in the give-and-take of resolving the dispute between these two companies. Vail reasserted the validity of the original patent, rejected a telegraph company's proposal to sell out to Western Union, refused to negotiate the sale of Bell's long distance services, and insisted on retaining control of local exchange telephone services. In fact, Vail took the offensive. He threatened to enter the telegraph market as a rival and competitor to Western Union.[5]

The result was perhaps surprising to onlookers and participants alike. Western Union engaged in a strategic retreat, acknowledged the validity of the Bell patent, withdrew from the telephone market, sold its patents to Bell, and transferred the American Speaking Telephone Company to Bell for $300,000. Bell, in turn, consented to pay Western Union 20 percent of future royalties and agreed to stay out of the telegraph business. By 1879, the telephone and telegraph markets in the United States were divided, if not cartelized. So victorious was the Bell Patent Association that the year 1879 is often referred to as the Magna Carta of the telephone industry.[6] Those fortunate enough to hold stock in the early Bell association soon found themselves financially independent.[7]

Bell's next step was twofold: to defend the patent and to enlarge the portfolio. The latter process was pursued in steps that Vail described as follows: "As quick as we started into the district exchange system we found out that it would develop a thousand and one little patents and inventions with which to do the business which was necessary and that was what we wanted to get control and get possession of."[8]

By 1896, Bell had purchased more than 20,000 patents related to all kinds of telephone equipment and apparatus. In the middle 1880s, telephone equipment included the wall set with a Blake transmitter, a lightning arrestor, a receiver with a hook switch, a pair of call bells, a small magneto operated by a crank and a battery on the floor of the customer's location, telephone wires, and crude manual switchboards.

Patent infringement suits were prosecuted vigorously. In the 1880s, some 600 suits were filed, 200 alone against the rival telephone company.[9] If a firm lost the suit, the equipment was publicly identified as an illicit intrusion. In Saint Louis, Bell put the torch to non-Bell switching equipment on a Mississippi levee as an object lesson.[10]

Vail was brilliant in both tactics and strategy, and he was equally forthright as to the purpose of his company's goal: "What we wanted to do was get possession of the field in such a way that patent or no patent we could control it."[11] But patents hold a finite life, and in ensuing years telephone competition sprouted and grew when the basic patents expired. At first, competition arose in rural areas and in areas Bell neglected to serve. But rivalry penetrated Bell's own territories and with rivalry came technical diversity. Some independent telephone companies offered automatic switching machines in contrast to Bell's manual switching offices. The first switch began in 1892 in La Porte, Indiana, a precursor to today's automatic switching services.

For some reasons, Bell missed this particular turn in technology and innovation. A Bell laboratory history suggests that Bell management felt subscribers did not want the burden of dialing their own phones, preferring to assign that function to the central operator. In any case, Bell found itself licensing switching technology from a non-Western Electric rival. As one official observed:

I can only recall one instance in which we have failed to purchase patents or inventions of that character and order and in this connection I refer to the automatic exchange patent. But as regards them we were reasonably sure that they never would be dangerous or valuable to competitors. In this, however, it has turned out that we were mistaken.[12]

In 1907, after an absence of nearly two decades, Vail returned to the parent telephone company, now known as AT&T, and began to put into place his corporate strategy. Prior to 1907, AT&T's research policy was to adapt ideas originating from university laboratories. Hammond V. Hayes, the director of Bell's mechanical department, described Bell's strategy as follows: "No one was employed who as an inventor is capable of originating new applications or novel design."[13] The threat of radio technology, however, was to alter this approach. Indeed, Hayes retired as director, and Bell's research in Chicago and Boston was consolidated in New York, perhaps originally as an effort to reduce costs. But Hayes's replacement, J. J. Carty, with uncommon prescience saw the potential of radio:

At the present time, scientists in Germany, France and in Italy and a number of able researchers in America are at work upon the problem of wireless telephony. While this branch of the art seems at the present rather remote in its prospects of success, a most powerful impetus would be given to it if a suitable telephone repeater were developed. Whoever can supply and control the necessary repeater will exert a dominant influence on the art of wireless telephony when it is developed.[14]

The vacuum tube, radio, posed as a threat to wire line telephony and its related investment. By 1910, Vail enlarged Bell's research and development effort beyond merely product adaptation and improvement. Vail sponsored basic research, an endeavor that literally pushed back the frontiers of telephone knowledge.

From a budget of $200,000 in 1910, basic research increased to $1.5 million before the onset of World War I and the number of personnel devoted to such research increased nearly fivefold.[15] The acquisition of the DeForest patent (1913) placed Bell in the center of radio development, so much so that by 1920 Bell was able to assert that it "possessed nearly all patents associated with radio."[16] Nevertheless, radio patent distribution was so dispersed that by World War I a stalemate had developed, broken only by the Signal Corps and Department of the Navy requirement to produce sets and tubes for the war effort.

But after the Armistice the patent stalemate resumed. No radio set could be fabricated without threat of legal infringement. In frustration, General Electric began negotiations with American Marconi to sell its Alexanderson transmitter. The problem was that American Marconi was a British firm.

The Department of the Navy, fearing that an offshore supplier would control the U.S. shipping communication apparatus, asked General Electric to create an entity in order to retain the basic patents within the United States. The result was the formation of a patent pool consisting of General Electric, Westinghouse, and AT&T. The new company was to be called RCA, the Radio Corporation of America.[17]

In 1924, the Bell System created Bell Telephone Laboratory (BTL) as a marriage between the staffs of Western Electric and AT&T headquarters. BTL thus institutionalized research, and the laboratory became essentially a patent factory. By the early 1930s, BTL was producing 6,000 patents a year and enjoyed a portfolio of 6,000 patents and licenses to 9,000 others.[18]

Bell's patents were both offensive and defensive, not unlike political spheres of influence. A memorandum in the late 1920s observed that Bell's research and development "shall maintain an active offensive in the no-man's land lying between it and potential competitive industries."[19]

By the start of World War II, the preeminence of Bell Telephone Laboratory was well established. Funded by a license contract upon operating company revenues, the breadth and range of BTL married basic development to applied product development. Western Electric, AT&T's manufacturing affiliate, paid for product development, and Bell's operating companies paid for basic development. In short, Vail had constructed a process to supersede an event. Research was institutionalized as an ongoing endeavor. It was to be a model later replicated by other American industries.

CUSTOMER STATION EQUIPMENT

In the early months, the telephone instrument was sold to the subscriber, but that policy was soon reversed. The instrument was leased only by the parent American Bell Company and then rented to the various affiliated Bell operating companies. The parent firm charged a license fee rental to each of the operating companies on the order of 4 percent.[20] When Bell acquired an independent telephone company, the telephone instrument was purchased and leased back to the subscriber.

In the late 1920s, AT&T sold the instrument to the operating companies, and the license contract fee was thereby reduced substantially. Some regulatory authorities had targeted the license fee as an excessive charge. Bell's operating companies in turn rented the station equipment to their subscribers. A service was provided rather than a phone sold. The concept of end-to-end service evolved that included manufacturing, repair, and telephone innovation. End-to-end service became a euphemism for monopoly.

The customer station apparatus of rival firms was not attached to the Bell network. If a user persisted in buying and owning his own phone, the customer forfeited the right to telephone service. And if a customer subscribed to both a Bell and a non-Bell telephone service, the denial of terminal attachment meant the possession of two phones. In Vail's world, a dual phone system constituted wasteful duplication and mandated the recognition that telephony was a natural monopoly.

INTERCONNECTION OF FACILITIES

AT&T, originally started as a long distance affiliate of American Bell, established long distance service between Boston and Providence under Vail's urging. Long distance technology was pushed to include loaded coils, mechanical repeaters, hard drawn wire, and phantom circuits. By the turn of the century, AT&T emerged as the dominant firm in the provision of long distance facilities and services in the United States. Inadequate financing prevented a challenge by an independent telephone consortium.

Bell's toll monopoly effectively isolated non-Bell independents. AT&T would not permit independent lines access to Bell long distance facilities through interconnection. Interconnection was granted to Bell operating company licensees only. Non-Bell firms were soon merged or amalgamated into the Bell operating companies' system. Bell introduced legislation canceling the franchises of accredited firms in its various districts. Bell's interconnection policy was so compelling that access was denied to independent companies even outside Bell's operating territory. Accordingly, competition at the local level attenuated.

Vail's concept of natural monopoly was to include telegraphy as well as telephony. By 1910, AT&T had purchased control of Western Union Company, and the results were impressive; cost reduction and night letters, money orders, and common billing practices were new innovations.[21] A threat by the Department of Justice prompted Bell to divest Western Union, to agree to interconnect to noncompetitive, non-Bell companies, and to cease Bell's acquisition policy.[22]

In the 1920s, Bell's work in the vacuum tube and radio found the company immersed in commercial broadcasting activities. AT&T's radio network stretched from New York to Nebraska and was linked by long distance telephone wires and facilities. Rival broadcasting chains sprouted up. Denied access to its Bell toll telephone facilities, they sought to lease the facilities of the Western Union Telegraph Company. Although transmission quality was oriented toward non-voice, it was not inconceivable that Western Union lines held the potential of a second long distance facility. In 1926, after a period of negotiations, AT&T withdrew from the broadcasting industry and sold its chain to RCA whose affiliate, the National Broadcasting Company, agreed to lease toll telephone facilities from AT&T.[23]

By World War II, Bell's interconnection policy was well established: no interconnection of tolls to rival or competitive local operating companies. Regulation was later to sanction what long lines policy had established as a matter of corporate practice. The U.S. domestic communication market in facilities and services had been split into neat demarcations: local telephone versus toll, telephone versus telegraph, telephone versus broadcast, telephone versus mail. These boundaries, imposed by industry or government, appeared to possess a legitimacy in fact. Each industry was acknowledged to possess unique traits, customs, and organizations.

MANUFACTURING

The common ownership of telephone service and manufacturing commenced at the beginning of the industry. In 1881 Bell, under the urging of Theodore Vail, acquired the largest communication supplier—Western Electric. In the process, Bell secured a source for its equipment and supplies, acquired telephone patents, and welcomed Western's engineering department. Western Electric received exclusive license to manufacture equipment and products under AT&T's patents and in effect enjoyed exclusive sales to the Bell operating companies' requirements. As Bell's patent portfolio enlarged, Western Electric's product line expanded correspondingly.

Western Electric refused to sell equipment to the independent telephone companies, a policy that led to the rise of non-Western Electric suppliers, particularly after the basic patents expired in 1894–1895. Western defended Bell patents with a vigor equal to that of its parent. From 1894 until 1910, Western launched a total of seventy-four infringement suits. That such complaints injected uncertainty into the market can be seen by the observation of one Bell member: "The policy of bringing suits for infringements on patents is an excellent one because it keeps concerns which attempt opposition to the new and exciting condition since it keeps them all the time changing machines."[24] At the same time, Western embarked on an acquisition process buying up competitive manufacturers of equipment and hardware. Both the Kellogg Company and Stromberg-Carlson, acquired by Western Electric, were subsequently struck down by state antitrust laws.[25]

When Vail returned to the Bell System in 1907, Western Electric was shaken up and reorganized. Operations were centralized, some 12,000 employees were let go, equipment was standardized, and new products were dropped. Vail even encouraged Western Electric to sell its products to independent non-Bell telephone companies.

Bell operating companies bought their equipment from Western Electric almost exclusively. Western Electric enjoyed access to AT&T specifications and engineering, and possessed privileged knowledge of Bell's construction programs. Not only was Western insulated from market rivalry in the traditional sense, but also it was protected from the uncertainty of its customers' short- and long-run demand. Given Western's market structure and access to patents, Western emerged as a manufacturing monopoly. AT&T insisted that Western's costs and prices reflected economies of size or scale, but no Bell operating company put that assertion to the test by soliciting competitive tender offering for equipment. The equipment stood foreclosed.

In the mid-1920s, Western Electric consolidated its markets and elected to confine its activities to the United States. A European division was sold to ITT on condition that the buyer would stay out of the United States for a period of fifteen years.[26] In 1930, Western Electric acquired the Teletype Corporation, thus expanding its product line into teleprinters.[27]

That Western Electric emerged as the predominant manufacturer was all but inevitable. Its ownership by AT&T, its access to patents, its basic R&D funded by Bell Telephone Laboratory, its access to specifications for equipment, its demand forecast, and its budget proposals of the Bell operating companies—all constituted a closed corporate family. The holding company family was obviously functional; Bell Laboratory designed equipment, Western Electric manufactured it, the Bell operating companies purchased hardware, and the subscriber enjoyed telephone service provided by a monopoly firm. Critics of vertical integration were reminded that the telephone company enjoyed economies of scale and that the nation's telephone service far exceeded that of any other country. "Have you made a telephone call in Paris?" became the rallying cry of those who defended holding company control of telephone supply.

The only problem was that monopoly—of any sort—has never enjoyed universal acclaim in the United States. That perception laid the foundation for Vail's next strategic masterpiece.

NOTES

1. Robert Conot, *A Streak of Luck* (New York: Simon and Schuster, 1978), pp. 82–83. Zenas F. Wilber was in debt to Major Bailey of Pollock and Bailey, employed by Hubbard as Bell's patent attorney. When Bell's and Gray's patent arrived the same day Wilber declared an interference:

"Major Bailey thereupon took Bell to see Wilber and convinced Wilber on cursory evidence that Bell's papers had been filed a short time before Gray's. Wilber dissolved the interference and gave precedence to Bell." Conot also notes: "Wilber let Bell examine all of Group papers and explained in detail Gray's proposed method of transmitting and receiving. Wilber then let Bell file an amendment to the application."

2. W. C. Langdon, "The Myths of Telephone History," *Bell Telephone Quarterly* (April 1933): 139. E. Gray wrote:

The history of the telephone will never be fully written; it is partly hidden away in twenty or thirty thousand pages of testimony and partly lying on the hearts and consciousness of a few whose lips are sealed, some in death and others by a golden clasp whose grip is even tighter.

See also Kenneth P. Todd, Jr., *A Capsule History of the Bell System* (New York: AT&T, 1972); and John Brooks, *Telephone, The First Hundred Years* (New York: Harper and Row, 1976), p. 48.

3. Horace Coon, *American Tel. & Tel.* (New York: Longmans, Green and Co., 1939), p. 53. Also see J. Warren Stehman, *The Financial History of the AT&T Company* (Boston: Houghton Mifflin, 1925), p. 27.

4. W. Rupert Maclaurin, *Invention and Innovation in the Radio Industry* (New York: Macmillan, 1949), p. 3. A telegraph official recalls: "We were telegraph men and did not think about alternative methods of communications."

5. Coon, *American Tel. & Tel.*, p. 58. See also Robert Sobel, *The Entrepreneurs* (New York: Weybright and Talley, 1974), p. 233.

6. Vail recalls: "The principal point in dispute was the interexchange telephone business now known as toll line or long distance. The Western Union Company wanted to protect its telegraph business. I wanted to make possible our dream of a universal service. We stood on that. The Western Union yielded." J. Leigh Walsh, "The Origin and Growth of the Telephone Industry in Connecticut," in *Connecticut Pioneers in Telephony* (New Haven, Conn.: Telephone Pioneers of America, 1950), p. 110. Also see Herbert N. Casson, *The History of the Telephone* (Chicago: A. C. Mclury, 1910), p. 84.

7. Albert Bigelow Paine, *In One Man's Life* (New York: Harper, 1921), p. 138.

8. FCC Special Investigation, Bell System's Patent Control: Its effect and suggested remedies, Docket No. 1, 1937, p. 10.

9. FCC, Investigation of the Telephone Industry in the United States, 76th Cong., 1st sess., House Document No. 340 (Washington, D.C.: U.S Government Printing Office, 1939), pp. 180–211 (hereafter referred to as FCC Telephone Investigation).

I have determined in the future to abandon this portion of the work (new) of this department, devoting all our attention to practical developments of instruments and apparatus. I think the theoretical work can be accomplished quite well and more economically by collaboration with the students of the Institute of Technology and probably Harvard College.(p. 183).

10. FCC Telephone Investigation, pp. 189–190.

11. Richard Gable, "The Early Competitive Era in Telephone Communications, 1893–1920," *Law and Contemporary Problems* 34 (Spring 1969): 347.

12. FCC Special Investigation, p. 10.

13. United States v. Western Electric Company, Civil Action No. 17-49 (DNJ 1949).

14. Coon, *American Tel. & Tel.*, p. 60.

15. Ithiel de Sola Pool, "Retrospective Technological Assessment of the Telephone Report to

the National Science Foundation,'' Vol. 1, *Research Program on Communications Policy, M.I.T.*, 1977, p. 37.

16. Coon, *American Tel. & Tel.*, p. 198.

17. L. S. Howeth, *History of Connections—Electronics in the United States Navy* (Washington, D.C.: Bureau of Ships and Office of Naval History, 1962), p. 376:

It was absolutely impossible to manufacture any kind of workable apparatus without using practically all the inventions which were then known. . . . That there was not a single company among those making radio sets for the Navy which possessed basic patents sufficient to enable them to supply without infringement . . . a complete transmitter.

18. United States v. Western Electric Company, Civil Action No. 17–49 (DNJ 1949).

19. ''It seems to be essential to the accomplishment of the AT&T Company's primary purpose . . . that it shall maintain an active offensive in the 'no man's land' lying between it and potentially competitive interest.'' (Brooks, *Telephone*, p. 181).

20. James McConnaughey and Manley Irwin, ''Rate Base Evaluation and Vertical Integration: Shifting Standards in Telephone Regulation,'' *Indiana Law Review* 54, No. 2 (1978–1979), p. 185.

21. Brooks, *Telephone*, p. 172.

22. FCC Telephone Investigation, p. 142. See also Brooks, *Telephone*, p. 135.

23. Brooks, *Telephone*, p. 171.

24. Coon, *American Tel. & Tel.*, p. 108.

25. U.S. Congress, House, *Consent Decree Program of the Department of Justice*, Hearings before the Antitrust Subcommittee (Subcommittee No. 5), Committee of the Judiciary, 85th Cong., 2d sess., 1958, Part 2, Vol. 1, American Telephone and Telegraph, p. 38. See also Harry B. Macmeal, *The Story of Independent Telephony* (Chicago: Independent Pioneer Telephone Association, 1939), p. 140.

26. FCC Telephone Investigation. See also Manley R. Irwin, ''The Communications Industry,'' in Walter Adams, ed., *Structure of American Industry*, 4th ed. (New York: Macmillan, 1971), p. 401.

27. Brooks, *Telephone*, pp. 187–188; see also ''Teletype,'' *Fortune* 5 (March 1932): 42.

3

REGULATION

Vail's goal, to "get possession of the field," was assured by the first decade of this century. But if a telephone monopoly was established, what or who would protect the consumer's interest? Could not a telephone monopoly burden the user with indifferent service and extortionate rates? The Granger movement in the United States suggested that imposition of public regulation on the railroad industry could serve as an institutional constraint upon naked corporate power. Should not then regulation be extended to telephone service and companies?

Originally, the Bell Company opposed regulation in any form, equating regulation at the state level with an assault upon the U.S. patent system. An early president of Bell put the case as follows:

Why should a telephone business be regulated as to price more than any other industry? The reasons offered are that telephone companies are monopolies that have been granted special privileges by localities. But as they are monopolies only by virtue of the patent system, but which is everywhere accepted, no state in fairness ought to destroy what the patent system has created.[1]

Indeed, when the state of Indiana legislatively extended regulation over Bell Telephone, Bell withdrew its service from the state. When the legislation was withdrawn, Bell reentered the state.[2]

Nevertheless, patents expired, and by the late 1890s, new firms entered and established telephone service not only in rural areas but also within Bell territories. Similarly, equipment manufacturers began to spring up, so much so that Western Electric refused to sell hardware to the non-Bell companies. By the turn of the century, the independent telephone industry had formed its own trade association and by 1910 accounted for nearly 50 percent of telephone stations in the United States.

Inevitably, too, independent telephone companies challenged Bell within its own markets. Competition so stimulated price rivalry that in Connecticut Bell's business party line rates dropped 20 percent and its residential party lines were cut by over 40 percent.[3] Bell introduced "flying squads" to enhance its marketing efforts and expedited innovation and new service with new features. Never before or after had the telephone industry experienced such rapid penetration of telephone service to the subscribing public, and never before had the industry ever witnessed such high rates of productivity.

Of course, competitive entry tended to reduce the telephone profits of the individual players and to dampen returns on investment. Bell's strategy—nonattachment of telephone equipment, noninterconnection of local or toll facilities, and a ban on Western sales to non-Bell companies—acted to impede competitive entry. Nevertheless, competition was very much a fact of life when Theodore Vail returned to the company in 1907. Vail was not only aware of market rivalry, but by the time he became AT&T's chairman Bell's debt had grown from $65 million to $200 million.

As Vail surveyed the environmental landscape and contemplated the future of the industry, he confronted three strategic options: competition, "postalization," and regulation. Clearly, competition was one option open to Vail. But as Vail viewed the industry, its requirements, service, and capitalization, he insisted that the industry was a natural monopoly and that open market rivalry was nothing short of industrial warfare.[4] Because the telephone constituted a natural monopoly, any attempt to inject market rivalry into this environment would result in duplication of plant and equipment which in his judgment was wasteful and inefficient. In contrast to an American firm where market competition yielded lower costs and prices, Vail claimed that quite the opposite would occur in the telephone industry. Thus, open competition was rejected as a viable option for the industry.

A second alternative invited nationalization or, as it was called in this country, "postalization." The U.S. Postal Service had always expressed an interest in acquiring control over the telephone industry. After all, it had operated the first telegraph system in the country. On occasion, bills introduced into Congress advocated passage of postalization of the telephone industry. The Postmaster General himself insisted that unified control of the mail and telephone would best serve the interest of the subscribing public.

During World War I, AT&T was indeed run by the U.S. Postal Service, but in 1919 the Senate agreed to release the telephone company to its shareholders.[5] Even so, as a former postal employee, Vail regarded postalization as simply a monopoly without accountability. Postalization constituted an unregulated economic power. Under this format, who would protect the public from any transgression of power, what would insure corporate efficiency and innovation, and who would guarantee responsiveness to the user public? Vail rejected the option of an unregulated monopoly.

A third alternative, private ownership coupled with public accountability, emerged as the model which Vail pursued. Regulation, as a superior policy alternative, would preserve the incentives of a private corporation and would protect the public interest in adequate service and reasonable rates. If a telephone company was to enjoy an exclusive franchise, a regulatory commission would ensure that the firm's conduct and actions comported with the public interest. That regulation would seal off market entry and competition was obvious. But that was not bad. As Vail observed: "Business courts (regulation) will soon bring order and security out of the present uncertainty and be a bulwark against future economic disturbance."[6]

Vail did more than borrow the concept of railroad regulation. He actively lobbied and marketed the extension of commission legislation. By 1910, both Robert La Follette in Wisconsin and Charles Evans Hughes in New York had extended the franchise regulatory concept to telephony.[7] Then, in 1910 Congress, in amending the Interstate Commerce Commission Act (Mann-Elkins), enlarged the Interstate Commerce Commission's jurisdiction over telephone and telegraph carriers. In 1920, Congress immunized telephone and telegraph carriers from antitrust attack if acquisitions were sanctioned by regulatory agencies.[8]

The Communication Act of 1934 reformalized the early legislative developments at both the state and federal level. The 1934 act created a separate regulatory agency, the Federal Communications Commission, embracing both broadcasting and common carrier. The commission was assigned telephone jurisdiction over rates, service facilities, and construction certificates.

Telephone regulation of Bell reached its height not in 1934 but in 1956, when AT&T agreed to confine its activities and services to regulated communication common-carrier activities.[9] In 1956, the Bell System consented to forego diversification outside of what then constituted telephony. In many respects, 1956 served as a capstone of Vail's strategic option that regulation stood superior to public enterprise or unfettered competition.

The Consent Decree with the Department of Justice ended an antitrust suit filed by the department in 1949. With much of its evidence coming from an FCC investigation in the 1930s, the complaint alleged that AT&T, by assigning its patents to Western Electric and by purchasing its products almost exclusively from Western, had extended its telephone monopoly to a manufacturing monopoly.[10] That extension, asserted the Department of Justice, violated the letter and spirit of the Sherman Antitrust Act.

The department's proposed remedy was harsh: a spinoff of Western Electric, a division of the company into three separate firms, and imposition of competitive buying practices upon the Bell System companies. In agreeing to confine itself to communication regulation, AT&T retained Western Electric, thus preserving Vail's model of vertical integration.

In 1956, the consent agreement appeared to be a strategic coup, an exercise in management foresight that retained market structure without any genuine corporate sacrifice. Although Bell's signing of the judgment stood as a tribute to management perspicacity, from the perspective of some Justice participants the decree was a stain on an otherwise commendable antitrust record. In any case, the 1956 Consent Judgment reinforced Bell's commitment to regulation and the public utility concept.

CONSEQUENCES OF REGULATION

Regulation embodied, approved, and sanctioned standard telephone company practices. Company policies became more than management decisions; they now possessed the force and power of the government. Violation of company practices transcended management decisions and provoked the police power of the state.

First, the subscriber did not and could not own and attach equipment to the dial-up telephone line. The telephone was company property, leased as part of an overall service commitment to the subscriber. Customer-owned equipment, known as "foreign attachments," would ultimately result in service degradation and compromise the universality of service to the public. In fact, some state utility commissions extended the foreign attachment ban to plastic covers slipped over telephone directories. Such covers, ruled state public utility commissions (PUCs), transgressed telephone company property.

Second, Bell's local companies refused to interconnect non-AT&T toll facilities. Alternatively, AT&T's long lines would not interconnect non-Bell local operating companies competing in a Bell territory. Regulatory bodies soon defended the validity of noninterconnection on grounds of service, performance, and common-carrier responsibility. Noninterconnection was formalized as filed regulatory tariffs.

Third, vertical integration of utility and manufacturer by definition limited access of non-Western Electric suppliers to Bell's equipment market. In owning both the buyer and seller of equipment, AT&T released specifications, controlled product allocations, and coordinated Western's output to the Bell operating companies.[11] If a purchase was consummated, Western would add its markup to the general trade product and bill the utility accordingly.

Occasionally, a state commission would inquire as to why a Bell operating company confined its purchases to Western Electric. The answer invariably turned on Western's economies of scale, large size, and lower cost, and that seemed to suffice. Even so, equipment prices entered a carrier's investment rate base and ultimately determined subscriber rates. Could state public utility bodies insure subscribers that equipment cost and prices were reasonable and prudent? Or were the Bell operating companies victimized by an equipment monopoly? State commissions could not answer that question with confidence, and so it was seldom asked.

The courts insisted that AT&T shoulder the burden of proving reasonable equipment prices. By the 1920s, AT&T supplied price comparison studies of Western and non-Western equipment. Invariably, Western's prices were deemed to be lower, thus justifying a sole source procurement by the Bell operating companies.[12] State PUCs, in accepting this study, could at least challenge any charge that equipment monopoly inflated a carrier's capital rate base. On the contrary, state PUCs cited AT&T's price comparison studies to verify product efficiency and product innovation. As a result, state regulatory agencies indirectly sanctioned a manufacturing cartel.

Fourth, telephone rate-making was reduced to an exercise in cost plus. A carrier was entitled to generate sufficient revenues to cover operating cost and to generate a return on capital investment (net). A utility would add up its expenses, revenues, and sales, and insert a markup on capital investment so as to equate total revenue with total cost.

The problem, of course, was that a fixed markup blunted any economic reward or penalty. If cost ratios became lax, the carrier could seek rate relief. On the other hand, if the firm was particularly cost effective, a commission might lower rates. Cost plus fixed return tended to anesthetize the carrot and stick of profits and losses. To the extent that utility commissions believed that economies of scale would rebound to the welfare of the rate payer, some economists speculated that regulatory lag was the source of progress and efficiency.

All the same, regulators believed that operating costs stretched over long time periods would yield low depreciation and low rates. Hence, PUCs imposed extended economic life on equipment so that annual depreciation expenses would be minimal in a capital-intensive industry. In some regulatory jurisdictions, equipment life was presumed to last up to forty years.[13]

Fifth, the cost plus calculus migrated to telephone manufacturing operations, transmitting a rate base mentality from utility to supplier. Western Electric, for example, would add up its cost and impose a markup on its equipment prices. Prices would then enter the carrier's rate base. That cost plus became part of the nomenclature of manufacturing can be seen by references to non-Western manufacturers as "cream skimmers."[14] Such nomenclature somehow implied that any integrated supplier possessed a silent franchise.

In any event, utility commissions, state and federal, remained ambivalent in evaluating equipment and manufacturing. Some states, if they detected exorbitant equipment prices, could disallow equipment cost into the carrier's rate base. But those actions were exceptional. Certainly the FCC never found occasion to disallow any of Western's prices to AT&T.

Finally, regulation invited, sought, and approved tiers of telephone rate subsidies. Over time, revenues were allocated so as to keep exchange rates to a minimum, and revenue subsidization evolved between city and rural, business and residential, toll and exchange. Nevertheless, rate structure issues—the relationship of particular prices to specific costs—were essentially deferred to company, management, and discretion. Commission regulation preferred to deal with aggregates: overall revenues, overall costs, overall rates of return.

The world of regulation viewed vertical integration and the public utility concept as eminently workable. The industry was perceived as progressive, rate base economics fair, natural monopoly a reality, surveillance an effective institution, investment manageable, and corporate practices subject to some regulatory oversight. Regulation proved to be a forum for consumer relief in a world endowed with, if not certainty, discretionary, technological change. Any challenge to economies of scale, telephone technology, exclusivity of service, monopoly of equipment, or even the threat to the premise of regulation was regarded as unthinkable, if not inconceivable. The only problem was that this world of static certitude was about to change.

NOTES

1. Horace Coon, *American Tel. & Tel.* (New York: Longmans, Green and Co., 1939), pp. 70–71.

2. Paul Latske, *A Fight with an Octopus* (Chicago: Telephone Publishing Co., 1906), p. 32.

3. J. Leigh Walsh, *Connecticut Pioneers in Telephony* (New Haven, Conn.: Telephone Pioneers of America, 1950), p. 203. "So low indeed were those rates running in several states from two to four dollars a month that they led to the loss for the first and last time of a connecting town from the company's area of operations." See also Walsh, *Connecticut Pioneers*, p. 227, and Richard Gable, "The Early Competitive Era in Telephone Communications, 1893–1920," *Law and Contemporary Problems* 34 (Spring 1969): 347.

4. John Brooks, *Telephone: The First Hundred Years* (New York: Harper and Row, 1975).

5. Ibid., p. 148: "Albert Sidney Burleson . . . would shortly make clear that he believed the Telephone and Telegraph Systems of the Nation should be 'postalized' . . . "

6. *AT&T Annual Report,* American Telephone and Telegraph, 1914, pp. 47–49.

7. "Vail supported the effort of Governor of New York, Charles Evans Hughes, Senator La Follette of Wisconsin who persuaded the state legislature to regulate utilities." See Annette Frey, "The Public Must Be Served," Part No. 59, *Bell Telephone Magazine* (March/June 1975): 5.

8. G. Hamilton Loeb, "The Communications Act Policy Toward Competition, A Failure to Communicate," *Duke Law Review,* No. 1 (March 1978): 15: "By clearing away the barriers to consolidation of competing systems the Willis-Graham Act announced a recognition that the provisions of telephone service facilities within a single exchange area was a natural monopoly which should be protected from the deleterious effect of competition."

9. U.S. Congress, House, *Consent Decree Program of the Department of Justice,* Hearings Before the Antitrust Subcommittee (Subcommittee No. 5), 85th Cong., 2d sess., Part 2, 1958, p. 35.

10. FCC Investigation of the Telephone Industry in the United States, 76th Cong., 1st sess., House Document No. 340, Washington, D.C.: U.S. Government Printing Office, 1939, pp. 189–190.

11. FCC, In the Matter of American Telephone and Telegraph Company, the Associated Bell System Companies, Charges for Interstate Service, AT&T Transmittal Nos. 10 939, 11027, 11657, Phase II, Trial Staff Testimony, M. R. Irwin, Docket no. 19129, 1975.

12. James McConnaughey and Manley Irwin, "Rate Base Evaluation and Vertical Integration: Shifting Standards in Telephone Regulation," *Indiana Law Review* 54, No. 2 (1978–1979): 185.

13. H. Hindin, "Large Scale Integration Latches on to the Phone System," *Electronics,* June 5, 1980, p. 114. M. Irwin and J. Ela,"Commission Regulation under Technological Stress: The Case of Information Services," *Public Utilities Fortnightly* 107, No. 13, June 18, 1981, p. 34. See also B. Perry, "Telco Equipment Depreciation Policies: Why Major Changes Are Needed—FAST!," *Telephony,* September 8, 1980, p. 89; and P. Schuten, "Digital Dawn for Telephony," *New York Times,* December 31, 1978, p. D5. "Typically a local switch depreciates over a period of 30 to 40 years and unlike small independents, Bell's operating companies for the most part are using very reliable analog equipment. 'Right now we have too big a capital investment to make such a conversion' Mr. Vigilante says."

14. FCC Docket No. 19129, Irwin testimony.

4

POLICY CHANGES

Policy shifts in terminal facilities, services, and integration proceeded slowly and hesitantly, only to gain momentum in the 1960s and 1970s. That change was particularly acute at the federal level—the FCC, the Executive Office of the President, the courts, and the Department of Justice. These agencies began to challenge the industry's longstanding policies. The ban on subscriber ownership of telephone sets was typical.

ATTACHMENT

In the mid-1950s, AT&T banned a cup-like device attached to the telephone instrument, and the FCC upheld the ban. Both the FCC and Bell insisted that the device would compromise the quality of the telephone instrument and transmission, as well as dilute authority over the system itself. But an appeals court overturned the validity of the Bell tariff with the observation that the attachment of the device was privately beneficial without being publicly detrimental.[1] In short, the user could own and attach this device to the telephone set. Although the Hush-a-phone device proved an exception, most customer equipment was banned from network attachment by both state and FCC tariff filings.

By the mid-1960s, the FCC launched an investigation into the growing use of computers tied to the telephone network.[2] In response to a question about terminal attachment, the computer industry was nearly unanimous in requesting the right to buy, own, and hook equipment to the dial-up network. These responses differed from past instances before the commission. In the past, attic inventors sought to link their security and other Rube Goldberg devices to the public network. Computer users, manufacturers, data processors, service bureaus, and customers, however, asserted their need for quality, speed, and reliability, a need that in some cases exceeded whatever the telephone industry could provide. These users in particular sought relief from the foreign attachment ban.

By 1968, the FCC ruled on the legality of both a device acoustically linked to the telephone network and a mobile radio system linked to a Gulf of Mexico oil rig. Bell confiscated the device. The firm, Carterphone, filed an antitrust complaint, and the case wound up before the FCC on grounds of primary juris-diction. In June 1968, the commission decided by a vote of eight to zero that customers could own and attach passive devices to the dial-up network beyond merely the Carterphone device.[3] Hence, a major telephone company policy of over eighty years had been breached. Even a presidential task force report implied that customer ownership of equipment could be attached, provided that adequate protection was insured for the network.[4]

INTERCONNECTION

Interconnection of private transmission facilities to those of the telephone industry, particularly those of the Bell System, had long been prohibited. In the late 1950s, the FCC upheld that interconnection restriction while permitting customers to construct and own private microwave facilities.[5] Then, in the early 1960s, the Communications Satellite Act introduced another source of satellite communication channels through the formation of an international consortium. But even in 1962 the domestic potential of satellite relay, though the subject of discussion and debate in Congress, was deemed in the industry to be unlikely in the foreseeable future.

But as early as the mid-1960s, domestic application of satellites seemed fea-sible: a network requested satellite relay, and the Ford Foundation pushed for the domestic application of communication satellites. A presidential task force advocated that COMSAT serve as a prototype to establish U.S. communications facilities via satellite, and by the early 1970s, the White House suggested that the FCC opt for an open skies approach.[6] In 1972, the commission announced and implemented a multiple entry policy for domestic satellite relay. For the first time, diversity in long-haul facilities was being pursued as a national policy through either the White House or the FCC.

Terrestrial microwave was also being explored in the 1960s. Microwave Com-munication Incorporated (MCI), a private firm, sought FCC common-carrier status to render and offer communications services between Chicago and Saint Louis, a service competitive with Bell's services provided by AT&T, General Telephone Electronics, and Western Union. At the time of its application, MCI proposed to violate the very essence of economies of scale of long distance facilities. But by the mid-1960s, the responses to the FCC's inquiry on the use of computers on telephone lines began to reveal some new information. Many responses sought alternatives to an analog long distance facility and network:[7] alterations in rate structure, different payment schemes, digital transmission, and switching, as well as other features oriented toward digital in contrast to voice transmission requirements.

In 1969, the FCC approved the MCI application by a narrow vote.[8] Inundated by other filings for common-carrier status, the commission in 1971 sanctioned the specialized carrier industry, thus competing with the private line services and facilities offered by AT&T's Long Lines Division.[9]

By the mid-1970s, the FCC ruled on applications for digital switching, satellite resellers, and shared-use facilities. The shared-use facilities, adopting technology funded essentially by the Department of Defense, leased communication facilities, attached computers as nose or switching devices, and provided a packet switching service in stark contrast to the analog circuit switching offered by the telephone industry. The FCC gave clearance to such services, thereby adding still another dimension to long-haul diversity. While the commission did segment market access, this policy represented a turnaround from the previous policy of regarding long-haul facilities as the essence of natural monopoly.

Although less dramatic, public policy began to perceive the vertical integration front as less than a closed issue. True, the 1956 Consent Decree effectively insulated AT&T and Western Electric from any external challenge. But when Congress began considering its communication satellites legislation in 1961, the ownership of utility and manufacturing became a critical issue of discussion and debate. In that controversy, the closed market of Bell and Western Electric became subject to analysis, question, and query as to whether that model would be moved to the satellite technology and its industries. The final legislation ordered COMSAT to buy all of its equipment, hardware, and supplies on a competitive bid basis, deferring to the FCC the implementation of procurement standards.[10] Subsequently, the commission imposed bidding requirements, not only on prime contractors to COMSAT's requirements, but also on lower tiers of subcontractors.

Although this policy was shortlived, Congress at least broached the vertical integration issue, namely, a regulated carrier securing its equipment and hardware from multiple and diverse sources of supply.

A presidential task force explored the Western Electric-Bell integration problem. That structural format had now become too generalized to set the industry norm: General Telephone and its equipment suppliers, Continental Telephone, United Telephone, and so on. After examining the pros and cons of the vertical integration issue, the task force report concluded that access to a closed vertical market was essentially a matter of antitrust.[11] Little did the task force realize how prophetic that observation would be.

If antitrust was regarded as unthinkable in the Bell-Western nexus after 1956, the integration of independents into the equipment market was nevertheless far from a resolved issue. General Telephone and Electronics (GTE), patterning itself after AT&T, acquired independent telephone companies and secured the ownership of switching, transmission, and other manufacturing affiliates. International Telephone and Telegraph (ITT), now a nonintegrated supplier in the United States, found its customer base continually contracting as GTE acquired independent telephone companies and merged them within the GTE family. ITT viewed its access to the Bell-Western Electric market as virtually impossible and sought its shrinking and declining share of the GTE market or the other independents. In the late 1960s, ITT filed an antitrust suit against GTE.

In the 1970s, ITT emerged victorious in its complaint.[12] The federal judge ruled that GTE should purchase its equipment on an arm's length basis from all subsidiaries and under equitable conditions. Although falling short of direct structural divestiture requested in ITT's original complaint, the GTE/ITT decision at least addressed the question of market foreclosure attendant upon vertical integration and attempted to achieve a remedy.

Even the FCC took another look at the controversial question of Bell's vertical integration. The investigation, begun in 1965, did not end until a decade later. A special FCC trial staff contended that competition had spurred Western Electric's performance, productivity, cost, price, innovation, and marketing efforts. The trial staff even insisted that competition in the equipment market had enhanced and expedited the economic performance of Bell Telephone Laboratory.[13] But more importantly the trial staff argued that vertical integration closed competitive markets, and it recommended that Congress be told of the benefits of open access in equipment manufacturing.

Policy then began to move away from total control of the telephone carriers, its production, services, and structure. Customer ownership of equipment, new entry into satellites, digital switching, packet switching, microwave communications, and hesitant moves into procurement suggested that the policy of adopting and implementing traditional carrier practices could no longer be counted on. The Bell System, joined by the independent telephone companies, strenuously opposed this drift of competition. Competition threatened revenues and control of their investment rate base, the longstanding rate structure, vertical integration, and manufacturing monopoly.

The industry was clearly upset. Its structure, practices, and policies, all of which had been approved by the commission, were now subject to criticism and attack at the federal level, as well as by telephone company unions. Market entry had also imposed a standard of productivity upon work rules and wage rates. When the Communications Workers of America (CWA) boycotted Holiday Inn for buying non-Bell telephone equipment, it was clear that organized labor had joined the industry in opposing competition for what they viewed as competition merely for the sake of competition.[14]

NOTES

1. Hush-a-phone Corporation v. American Telephone and Telegraph Company, 238 f 2d 266 (D.C. cir—1956).

2. FCC, In the Matter of Regulatory and Policy Problems Presented by the Interdependence of Computers and Communications Services Facilities, Docket No. 16979, *Final Decision and Order*, March 1971.

3. FCC, In the Matter of Use of the Carterphone Device in Message Toll Telephone Service; In the Matter of Thomas F. Carter and Carter Electronics Corporation, Dallas, Texas, Complainants v. American Telephone and Telegraph Company, Associated Bell System Companies, Southwestern Bell Telephone Company, and General Telephone Company of the Southwest, Dockets No. 16942 and 17073, *Final Decision*, 1968.

4. President's Task Force on Communications Policy, *Final Report*, U.S. Government Printing Office, December 7, 1968, p. 28:

In response to the Commission's decision, the telephone companies have filed new tariffs to permit use of customer owned terminal devices and interconnection of private systems, subject to a number of limitations which the telephone companies say are necessary to protect the integrity of the switched network. These limitations would require the customer to comply with technical specifications set forth in the tariff; and to use a protective interface device and a network control signalling unit supplied by the telephone company.

These protective measures, some of which are still under challenge before the commission, would appear to reduce in importance the issue of "system integrity" which has long been the basis for excluding private equipment and systems from use with the switched network. If so, the path would be cleared for the development and use over the switched network of a wider range of terminal devices (particularly specialized terminal equipment necessary for transmitting data) and perhaps also of private communications systems.

5. FCC, In the Matter of Allocation of Frequencies in the Bands Above 890 Megacycles, Docket No. 11866, *Memorandum, Opinion and Order*, September 1960.

6. Letter to Dean Birch, Chairman, FCC, from Peter Flanagan, White House, January 30, 1970, Docket No. 16495, Appendix B, p. 23.

7. S. Mathison and P. Walker, *Computers and Telecommunications: Issues in Public Policy* (Englewood Cliffs, N.J.: Prentice-Hall, 1970), Appendix C, pp. 242–243.

8. FCC Application of MCI, Incorporated for Construction Permits to Establish New Facilities in the Domestic Public Point to Point Microwave Radio Service in Chicago, Illinois, Saint Louis, Missouri, and Intermediate Points, Docket No. 10509, 1969.

9. FCC, In the Matter of Establishment of Policies and Procedures for Consideration of Application to Provide Specialized Carrier Services in the Domestic Public Point to Point Microwave Radio Service and Proposed Amendments to Parts 21, 43, and 61 of the Commission's Rules, Docket No. 18920, *First Report and Order*, June 3, 1971.

10. FCC, In the Matter of Amendment of Part 25 of the Commission's Rules and Regulations with Respect to the Procurement of Apparatus, Equipment, and Services Required for the Establishment and Operation of the Communication Satellite System and Satellite Terminal Stations, Docket No. 15123, *Report and Order*, April 3, 1964.

11. President's Task Force on Communication Policy, *Final Report*, p. 40: "It is of course outside our competence to express judgment on the AT&T/Western Electric ties regarding the consistency with the Antitrust laws. That issue must be left to the Justice Department and to the Courts."

12. International Telephone and Telegraph Corporation v. General Telephone and Electronics Corporation and Hawaiian Telephone Company, Civil Action No. 2754, U.S. District Court, Hawaii, 1969. See also ITT Corporation v. GTE Corporation, 528FD. 913 (9th Circuit), 1975.

13. FCC, In the Matter of American Telephone and Telegraph Company, the Associated Bell System Companies, Charges for Interstate Telephone Service, AT&T Transmittal Nos. 10989, 11027, 11657, Docket No. 19129 (Phase II), *Final Decision and Order*, March 1977, p. 52. See also Manley R. Irwin Testimony in Docket No. 19129 (Phase II), Trial Staff Exhibit 172; Frank Barbetta, "AT&T Marketing Realignment Plant," *Electronic News*, February 27, 1978, p. 1; "Is That You, Ma Bell?," *Sales Management*, Vol. 115, no. 8, March 3, 1975, pp. 30-34. "Prior to competition we introduced things when we were good and ready." "A Clear Path for Those In Sales and Other Business," *Sales Management*, March 3, 1975; and "New Muscle in Marketing" *The Bell Telephone Magazine* (July-August 1974): 20.

14. U.S. Congress, Senate, Industrial Reorganization Act, Hearings before the Subcommittee on Antitrust and Monopoly, Committee on the Judiciary, 93d Cong., 2d sess., Part 5, The Communication Industry, June 1974, pp. 2972-2974: "The Holiday in Ridge Road went Interconnect last week. No more parties or meetings at the Holiday Inn West; let the Interconnect company have them" (CWA Local 1170).

5

BELL'S COMPETITIVE RESPONSE

However tentative, market rivalry in terminals, facilities, and equipment presented an unprecedented challenge to AT&T. Market entry established a standard against which AT&T's economic performance could be assessed. Bell System's response was twofold: an internal management reevaluation and an external adjustment to a new environment of rivalry. The customer terminal equipment market was first to experience AT&T's response.

TERMINALS

AT&T's adjustment to competitors' entry in customer station equipment included a reexamination of its research, manufacturing, operating companies' policies, and holding company strategy. In view of the additional choices conferred on the subscriber, the Bell System was confronted with a benchmark that measured its performance on a broad scale. AT&T discovered, for example, that some of its product line was obsolete.[1] In other cases, Bell acknowledged a gap in its PBX (Private Branch Exchange) or key system product line.[2] In still other instances, Bell's PBX and key telephone product costs were out of line, features were missing, and maintenance installation cost did not compare favorably with that imposed by its interconnect competitors.[3] As one operating company commented on the impact of rivalry: "The competition we are experiencing is really symptomatic of us not providing the proper products and services to meet our customer needs."[4]

This realization prompted Bell to reorient its R&D and to expedite product innovation.[5] Western Electric reexamined its own product line, production cost, and product prices billed to the Bell operating companies.[6] The operating companies introduced flexible pricing programs, reorganized their marketing departments, and targeted key telephone or PBX systems to subscribers evaluating competitive alternatives.

To the extent that market competition enhanced corporate innovation, productivity, and responsiveness, the subscriber obviously gained in product choice, features, and installation time. On the other hand, Bell's response to interconnect competitors was singularly controversial. Interconnect suppliers asserted that the operating companies dropped tariffs below their fully allocated costs, packaged lines and terminals as a communication tie-in, and mandated that subscribers rent a protective device if they elected to buy equipment on the open market. Bell opposed the direct attachment of customer products to their lines and opposed an FCC certification program proposing such attachment.[7]

Interconnect suppliers insisted that posting full cost prices on telephone monopoly services but employing incremental pricing for competitive services constituted cross-subsidization with its attendant effects on market competition. Interconnect firms argued that Bell's protective device inserted between a customer's phone and Bell's line was discriminatory and anticompetitive. No such device was required for Bell's equipment—even though in some cases the equipment was identical to interconnect products.

Bell's protective device upped the cost of buying equipment and placed the interconnect supplier at a competitive disadvantage. Moreover, interconnect suppliers contended that Bell's delivery of the required protective device was sometimes tardy and that the coupler device was occasionally defective. AT&T insisted that the protective device was designed to eliminate potential harm to the telephone system.

The interconnect industry was particularly frustrated by AT&T's opposition to the FCC's certification program.[8] Had such a program been adopted, it would have eliminated Bell's coupler interface device. AT&T insisted, however, that a protective device was important for the compelling reason that the company bore ultimate responsibility for the quality and integrity of a nationwide telephone system.

And AT&T insisted that the FCC's certification procedure did not lend itself to easy resolution, given the complexity of equipment standardization. Indeed, AT&T resisted the concept of direct attachment of user terminals to the dial-up network. After nearly four years of study, the FCC requested AT&T to issue technical standards for non-Western Electric equipment. By late 1975, Bell's specifications became available.

The commission, however, did more than accept AT&T's technical standards. The FCC extended the requirements to include Western Electric's equipment as well. Bell's standards were to apply to its own telephone equipment. It was then that Bell informed the commission that its products fell short of the specifications imposed on competitors' equipment.[9] Within a six-week period Bell issued a modified set of performance specs to the FCC; and then Bell took its opposition to the courts.

AT&T's response to interconnect competition chilled the environment of these new firms. Price adjustments, tariff restrictions, interface devices, and certification process posed significant disincentives to existing and prospective firms. Some firms survived but withdrew from the interconnect field. A rash of antitrust suits soon alleged that Bell's competitive response had foreclosed market entry, limited rivalry, discouraged choice, and reinforced AT&T's monopoly of telephone equipment. Some suits were settled out of court; others vindicated Bell's actions. On the other hand, a jury awarded a $276 million antitrust suit to Litton Industry on the grounds that Bell's conduct had violated the antitrust status, a decision upheld on appeal.[10]

FACILITIES/SERVICE

In selling R&D equipment, hardware, and labor, specialized carriers obviously posed a challenge to AT&T's long distance facilities and services. AT&T's response was far reaching; research and development was expedited, manufactured products were introduced, new tariffs and costing principles inaugurated, marketing and management reorganized, and product features and services enhanced.[11] Whether measured in terms of efficiency, innovation, or user responsiveness, competition served as a spur to economic performance.

Bell's reaction to specialized carrier rivalry was controversial. When AT&T dropped private lease rates, in some cases up to 85 percent of former levels, specialized carriers charged that such reductions were not matched by comparable reductions in costs.[12] Specialized carriers asserted that Bell's ability to employ fully allocated pricing for monopoly markets and to engage in incremental pricing for competitive markets marked the essence of cross-subsidization. Such cross-subsidy, they asserted, served to preserve AT&T's monopoly of interstate lease services and facilities.

Bell's interconnection policy also proved a contentious issue. AT&T originally denied specialized carriers access to local distribution facilities outright. Later, Bell granted interconnection for certain leased or dedicated services. In still other instances, AT&T informed the specialized carriers that they must purchase local access facilities. Finally, Bell operating companies restricted local distribution facilities to what they defined as private line services only.[13]

Specialized carriers insisted that local Bell companies favored AT&T's long lines over its interstate rivals in terms of loop access and dialing codes. Specialized carriers argued that AT&T granted interconnection to Western Union while denying similar access to specialized carriers. Bell responded that its tariffs had been approved by state regulatory commissions, that the premise of incremental pricing represented a valid pricing mechanism, and that its competitive services constituted a legitimate response to market competition and rivalry.

AT&T also maintained that the FCC had not contemplated that competitive firms offer CCSA (Common Control Switching Arrangement) and FX services (private switched services). According to Bell, the FCC had not intended to introduce rivalry into toll switched telephone services such as MCI's Execunet Service. Rather, the FCC limited competition to point to point leased communication services. The FCC, in fact, supported Bell's contention and ordered AT&T to deny MCI access to local distribution facilities.[14]

Whatever the merit of AT&T's position, its reaction to outside competition signalled a spirited opposition to competition in any form. Specialized carrier revenues were postponed, legal costs increased, and return on investment further postponed. As regulatory due process examined the intricacies of fully allocated pricing versus long-run incremental pricing precious time was obviously consumed. Some specialized carriers withdrew from the market (Datran); others merged and consolidated. The survivors, ITT, Southern Pacific, and MCI, eventually filed antitrust suits alleging that Bell's market conduct was predatory and anticompetitive.[15] A Chicago jury awarded MCI nearly $2 billion in damages. An appeals court upheld AT&T's liability, but sent the case back to the district court for a retrial on damages. On the other hand, a Washington, D.C., district court judge dismissed Southern Pacific's antitrust complaint against AT&T on grounds that Bell as a public utility was subject to both state and FCC regulation.[16]

EQUIPMENT

Telephone manufacturing, long characterized by vertical integration, preordained that Bell purchase equipment, transmission, terminals, and switching from Western Electric. To the extent that buying and selling took place within a single corporate family, general trade suppliers were generally regarded as outside and beyond the corporate family.

Given vertical integration, competition in the procurement and supply and manufacturing of equipment was essentially nonexistent. The holding company laboratory designed products, Western manufactured equipment, the operating companies bought equipment incident to their service offerings to the public. By the late 1960s, however, general trade suppliers began to make incursions into isolated product submarkets. This limited entry occurred despite vertical integration and despite a policy of essentially sole-source procurement with the holding company family.

AT&T, encountering competition in equipment manufacturing, undertook an examination of the cost, price features, technology, delivery time, and product availability. General trade competition stimulated the BTL to expedite its development process, to reorder research priorities, and to develop direct contact with the various Bell operating companies (traditionally banned). Competitive suppliers prompted Western Electric to reexamine its product cost, price, features, and technical progressivity.

Western Electric acknowledged that the Bell operating companies were delighted with Western's new attitude: "They [Bell Operating Telephone Companies] are now starting to say Western has been a big help to them, that Western has come alive."[17] On the other hand, outside suppliers insisted that AT&T continued to favor Western Electric equipment. These suppliers argued that Bell delayed releasing equipment specifications to competitors, bought outside equipment only as a temporary gap filler, and secured products from Western Electric in spite of the lower prices available in a competitive market.[18]

General trade suppliers further contended that Western Electric's crash product development programs were motivated by intent to foreclose competitive markets, that operating companies in some instances ordered Western equipment without so much as a price quote, and that Western as a supplier of equipment for operating companies was hardly in a position of detached objectivity.[19]

AT&T insisted that vertical integration accorded the operating company product quality, lower prices, and enhanced features. The Bell System repeatedly stated that it was driven to secure the best equipment at the lowest price. By the mid-1970s, AT&T in fact established a new organization to assist the Bell operating companies in buying and making available non-Western equipment from outside suppliers—a unit called the Bell Systems Purchase Product Division.[20]

Nevertheless, several manufacturers filed antitrust suits alleging that AT&T's integration combined with operating company buying practices froze effective competition out of the equipment market. In some cases, the suits were settled out of court (ITT, Telesciences); other cases remain pending.[21]

The rash of antitrust suits in terminals, facilities, equipment switching, and transmission obviously proved disconcerting to a regulated utility. Bell's policies were subject to the jurisdiction of the state and federal agencies and its tariffs were presumably sanctioned by utility commissions. Nevertheless, AT&T was beset by some forty private antitrust suits, each alleging in one way or another that Bell's structure and conduct was predatory and anticompetitive.

As if these private suits were not enough, the Department of Justice filed an antitrust complaint in November 1974 charging that AT&T had monopolized the subscriber equipment market (interconnect), foreclosed market entry into intercity services (specialized carriers), and had preempted access to equipment manufacturing.[22] The Department of Justice insisted that nothing short of a massive corporate restructure would suffice to restore competition in each of these communication submarkets.

In 1976, AT&T sponsored legislation that sought to immunize the company from market entry and antitrust liability.[23] Indeed, the legislation contemplated a restoration of its monopoly status in terminals, facilities, and manufacturing. The legislation, which proposed shifting regulatory jurisdiction of terminal equipment to state commissions, essentially proposed a moratorium upon market entry and sanctioned the acquisition by AT&T of its interstate communication competitors. The bill, the Consumer Communication Reform Act, was premised on preserving the transfer of a toll subsidy to local exchange rates so as to make residential telephone rates affordable. The bill did not pass.

In the meantime, AT&T's antitrust difficulties were multiplying. Bell now found itself confronting the power and purse of the federal government. By 1982, Bell acknowledged that it had spent some $1 billion in antitrust expenses.[24] While legal fees are allowable for rate-making purposes, antitrust was consuming management's time, energy, and resources. And, worst of all, the judge at the federal trial gave every indication the government was making a compelling showing.

NOTES

1. United States v. America Telegraph and Telephone Company, et al. U.S. District Court, District of Columbia, Plaintiffs Third Statement of Contentions and Proof, January 10, 1980, p.

1313 (hereafter cited as Government Antitrust Suit). See also, FCC, In the Matter of American Telephone and Telegraph Company, the Associated Bell System Companies, Charges for Interstate Telephone Service, AT&T Transmittal Nos. 10989, 11027, 11657, Docket No. 19129, Phase II, FCC Trial Staff Submission, 1975, p. 502 (hereafter cited as Docket No. 19129, Phase II).

2. Docket No. 19129, Phase II, p. 393.

3. "Bell Labs Answers Calls for Help," *Business Week,* January 23, 1971, p. 38. See also Docket No. 19129, Phase II, Trial Staff, Exhibit 146, November 1974, pp. 478, 484.

4. "Pacific Telephones' Product and Service Manager Approach to Competition," *Telephone Engineering and Management,* December 15, 1973, p. 52.

5. FCC, Docket No. 19129, Phase II, Exhibit, Corporate Planning Report AT&T, 1972; Trial Staff Exhibit 147, 1974. See also Trial Staff's Brief, February 2, 1976, p. 42.

Several examples illustrate the urgency of product innovation:

- the 77A was developed on a crash program basis (PFD 203)

- the 208 was to be introduced "as rapidly as possible" (PFD 16)

- 812A development was initiated on an expedited basis (PFD 231)

- the Comkey 718/1434 were developed on a "short fuse schedule"(PFD 45)

- the 801A development was on an "expedited basis" (PFD180)

- the ACD-CO was a "very important and very short schedule project" (PFD 99)

- the 802 hotel/motel PBX was developed with a sense of "competitive urgency" (PFD 184)

6. Government Antitrust Suit, pp. 1102–1103.

7. M. J. Richter, "Ex-Bell Chairman: We Thought Competition Would Lower Quality," *Electronic News,* August 31, 1981, p. 21: "Even if there had been wonderful standards at the time, you would have opposed certification because of open competition," asked the judge. "Yes," replied Mr. deButts (AT&T chairman).

8. J. Fraser, "AT&T's de Butts Blasts Interconnects," *Electronic News,* October 16, 1972. The AT&T chairman stated, "I agree that many people can make good equipment as Western Electric. The big problem is maintenance." (p. 17).

9. "Standards Eased on Gear Added to Phone Network," *Wall Street Journal,* March 17, 1976, p. 20: "The initial technical standards for the registration program were based primarily on specifications AT&T had set for its competition in the equipment manufacturing business, according to an FCC staff member. However he explained when the Commission said that AT&T's own equipment also would have to meet those standards the phone company told the FCC they were too tough."

10. D. Mancuso and F. Barbetta, "Litton Awarded $276 M in Suit Against AT&T," *Electronic News,* July 6, 1981, p. 1.

11. S. E. Bonsack, "A Discussion of Marketing in Telecommunications Under Conditions of Growing Competition," in *New Challenges to Public Utilities Management,* ed. H. Trebing, Institute of Public Utilities, Michigan State University (East Lansing: Michigan State University Press, 1974).

12. FCC, Report of the Telephone and Telegraph Committee of the FCC in the Domestic Telegraph Investigation, Docket No. 14650, 1965, p. 181.

13. R. Vinton, "MCI Charges AT&T Tried to Put It Out of Business," *Electronic News,* February 11, 1979, p. 1: "A key element of Mr. McGowen's testimony last week however was MCI's charge that AT&T at first refused to provide necessary interconnections to set up competitive private line long haul services. And when AT&T did provide MCI with the interconnection it made them restrictive and overcharged the smaller firms, MCI."

14. FCC, In the Matter of MCI Telecommunications Corporation, Investigation into the Lawfulness of Tariff FCC No. 1 insofar as it purports to offer Execunet Service, Docket No. 20648, *Decision,* July 13, 1976.

15. Southern Pacific Communications Company, et al. v. American Telephone and Telegraph

Company, et al., Civil Action No. 78–0545, U.S. District Court, District of Columbia, Plaintiffs memorandum in Opposition to Defendants' Motion for Involuntary Dismissal under Rule 41B, June 15, 1982. See also "AT&T Settles Datran Suit for $50 M," *Electronic News,* March 17, 1981, p. 1.

16. R. Frank, "A Big Blow to AT&T," *Datamation* (August 1980): 57. Also see Southern Pacific Communications Company, et al. v. American Telephone and Telegraph Company, et al., Civil Action No. 78–0545, U.S. District Court, District of Columbia, Memorandum Opinion of U.S. District Judge C. R. Richey, December 21, 1981.

17. FCC, Docket No. 19129, Phase II, Sales Service Consulting Newsletter, South Western Region, letter from W. J. O'Donnell to J. F. McQuarrie, December 7, 1972, p. 2. See also FCC, In the Matter of Bell Operating Company Procurement of Telecommunications Equipment, Petition of ITT, May 30, 1979, p. 15 (hereafter cited as ITT Procurement Petition): "In order to retain this business for Western, BTL accelerated development of D-3, Western began to manufacture D-3 from preliminary design drawings." See also FCC, Trial Staff Proposed Findings of Fact and Conclusion, 1976, p. 523. "Rather than purchase the NA409 the Customer Products Council approved the immediate development of a product of similar design because they felt it 'mandatory' that this project be manufactured by Western Electric." See, finally, FCC, Docket No. 19129, Phase II, Testimony of M. R. Irwin, Trial Staff Exhibit, 1975, p. 24.

18. ITT Procurement Petition, p. 5: "Western devised unique installation procedures which were unacceptable to the operating company, and New York Telephone agreed to slip its installation dates and to use D-3 without a field trial and without a price (ITT Proposed Findings, pp. 15–20). An Executive Vice President of BTL gloated to Western Electric personnel that through the efforts of the task force, which he described as 'fancy footwork', the matter had been resolved to 'everyone's satisfaction (everybody, except ITT)."

19. ITT Procurement Petition, p. 15.

20. M. J. Richter, "ITT Charges Bell WE Plan Violates Arm's Length Rule," *Electronic News,* July 30, 1979, p. 21; see also ITT Procurement Petition. CONRAC Corporation v. American Telephone and Telegraph Company, Telesciences Inc., Bell Telephone Company of Nevada, et al., U.S. District Court, Southern District of New York, April 13, 1982; Telesciences Inc. v. American Telephone and Telegraph Company, Bell Telephone of Nevada, et al., U.S. District Court, District of Columbia, September 25, 1980 (settled out of court).

21. International Telephone and Telegraph Corporation v. American Telephone and Telegraph Company, Western Electric, Inc. and Bell Telephone Laboratories, Inc., U.S. District Court of New York, June 10, 1977 (settled February 28, 1980). AT&T agreed to the possible purchase of $2 billion in telecommunications products over a ten-year period.

22. United States v. American Telephone and Telegraph Company, Western Electric Company, Inc.; and Bell Telephone Laboratories, Inc.; and Government Antitrust Suit.

23. "Communications Dogfight," *Dun's Review* (June 1977): 48.

24. "Final Shape of Big AT&T Settlement," *U.S. News and World Report,* August 30, 1982, p. 36. See also S. Aug, "AT&T Puts Cost of U.S. Antitrust Suit at $1 Billion," *Washington Star,* February 17, 1978, p. 20.

6

TECHNOLOGY AND MARKET ENTRY

By the middle 1960s, it was evident that the pace of technology had begun to accelerate. As an engine to market entry, technology disrupted the quiet universe of the regulated firm and challenged historic patterns of prices, costs, depreciation, write-offs, marketing, and innovation. What was the reach of technology? How did it abet market entry? And what was the effect of the entry process on a firm's boundaries?

TECHNOLOGY

Microelectronic technology embodies a cross-current of forces. It embraces a range of disciplines, a family of technology that yields products, and features and complexity that reduce equipment size and bulk.

A first trait of technology is its diversity—optics, mathematics, molecular chemistry, software, electrical engineering, astronomy, particle physics, microbiology—in short, a family of disciplines. The range of technology suggests that a heterogeneous mix of disciplines now assumes importance. One computer manufacturer asserts it is pursuing forty to fifty technologies at the same time.[1]

Second, technologies interact, couple, and combine. It is a cliché to observe that computers and communications converge, a confluence driven by advances in solid state technology from the transistor, the integrated circuit, the microprocessor, and the VLSI (very large scale integration) (Figure 6.1).[2] But that coupling of computers and communications is indicative of an interdependence between satellites and computers, fiber optics and aerospace, telephones and software.[3]

Third, technology fosters productivity. The number of components imbedded in a silicon chip has doubled annually over the past twenty years. Recently, a 32 bit microprocessor embodying 450,000 transistors per chip was announced, but within months another product boosted the density to 600,000 transistors per chip. The target of 1 million elements per chip is now regarded as eminently within reach.

Chip density yields still another bonus. Products once weighing tons or pounds

Figure 6.1
Perspective of "C&C"

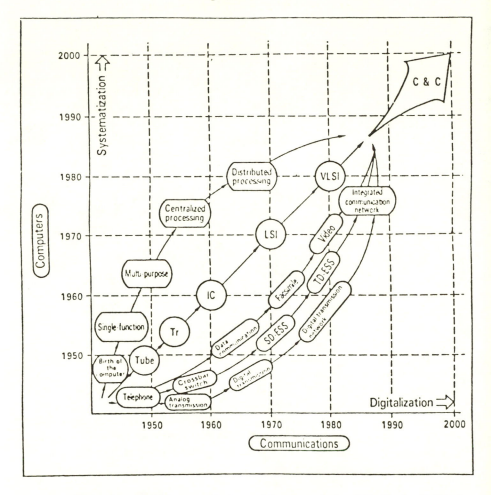

Source: Koji Kobayashi, "Telecommunications and Computers: An Inevitable Marriage", *Telephony*, January 28, 1980, p. 78.

are reduced to ounces; this trend has persisted for over a generation. Products once stored in a room can now be carried in one's pocket. A digital watch can double as a telephone or a calculator. Obviously, miniaturization alters one's perception of optimal plant size or scale economics.

Fourth, chip complexity yields another trait—a multiplication of features. A product's function today may be imbedded in a component tomorrow. And product features proliferate and multiply. A telephone yesterday may activate a computer, record time, and store messages today. Tomorrow the features of a

Table 6.1
AN INFORMATION INFRASTRUCTURE

Manufacturing Terminals	Networks	Services
Telephone set	Satellite	Computer-aided design
Telegraph set	Fiber optics	Computer-aided manufacturing
PBX	Pair copper wire	On-line robots
Minicomputer	Coaxial cable	On-line automatic testing
Mainframe computer	Electric power line	equipment
Personal computer	Microwave radio	Word processing
TV set	FM broadcast	Electronic funds transfer
FM radio set	AM broadcast	Remote data base
VCR	Direct broadcast satellite	Remote processing
CRT set	Low power satellite	Electronic mail
Work station	Infrared transmission	Voice mail
Copier	Digital broadcast	Teleconferencing
Electronic typewriter	Packet switching	Satellite scanning
Fax machine	Analogue switching	On-line software
Satellite earth terminal	Digital switching	Home information services
Paging unit	Data concentrators	Videotext
Graphic terminal	Local area networks	Teletext
Robot		Pay TV
Point of sale terminal		Cable TV
		DDP
		Remote paging
		Sensoring services

telephone and computer will be indistinguishable. Indeed, Atari, a manufacturer of computer games, has announced its entry into the telephone equipment market.

Finally, research activity is hardly confined to telecommunications and computers. Some two dozen industries in the United States directly or indirectly participate in some facet of R&D. To this private endeavor must be added the resources of the government or public sector.[4]

It is also clear that information technology is now of global dimension. Whatever the relationship between government and the private sector, international resources devoted to advances in the state of the microelectronic art emphasize that the technological genie is out of the bottle. This global investment in research microelectronics, aerospace, software, and optics in turn produces a network of terminals, communications, networks, and services—in a word an information matrix.[5] (Table 6.1.)

AN INFORMATION INFRASTRUCTURE

An information infrastructure provides a focus for the dynamics of the entry process. Within the matrix format, a firm may move horizontally, vertically, or in both directions simultaneously. A firm, for example, may begin by manufacturing IC chips and then evolve into telephone terminals and switching equipment manufacturing. The Mitel Corporation illustrates this horizontal transformation from component to system supplier.[6]

A firm, originating as a network provider, may integrate backward into the supplying and manufacturing of terminals and component fabrication. Simultaneously, the firm may expand the range of services by moving down the service column. COMSAT, which illustrates this process, began as a satellite carrier, diversified into chip manufacturing, and is expanding into electronic mail, teleconferencing, and satellite scanning.[7] Similarly, a local telephone company by offering essentially a local exchange service may diversify into software consulting services, long distance telephone services, and office equipment or automation equipment (Rochester Telephone).[8]

A firm may start in the service sector and diversify horizontally into networks and terminals. Tymeshare, for example, originally offered a payroll data processing industry administrative service to its remote customers.[9] The company has become a value-added carrier and via a joint venture provides intelligent terminals to its customers.

A firm may start as a value-added carrier and through digital termination systems (radio techniques) bypass local telephone distribution facilities. Office automation may lead a firm into remote communication services, teleconferencing services, electronic mail, and data processing to offices, banks, and a variety of other customers, a diversification exemplified by IBM.[10]

A firm may plug the information matrix cells by pursuing a row/column diversification simultaneously. Mitel, Inc., by moving from components, terminals, and satellite dishes into teleconferencing, data, and image processing services (through a joint venture with American Satellite)[11] illustrates this pattern.

Finally, a firm offering voice services may diversify into terminal and networks vertically. AT&T's activities in videotext data base systems, intelligent terminals, and loop bypass techniques suggest this strategy.[12]

THE DRIVING FORCE

What are the economic forces behind market entry, and will these forces persist in the foreseeable future? An inquiry into the first question leads to an examination of unit costs and projected revenues. Table 6.2 suggests that the unit costs of a variety of products—from satellite dishes to fiber optics—have declined in the past fifteen to twenty years. As costs translate into falling unit prices, capital investment as an economic barrier recedes and declines over time. Over the past twenty years, electronic memory costs have declined 40 percent annually, computer logic costs 20 percent annually, and communication costs 11 percent annually.[13]

If one surveys the projected costs of microprocessors, terminals, and memory, there is every reason to believe that these products will experience even greater productivity gains in the future. Industry comments suggest that the microelectronics revolution has yet to run out of steam:

- The microelectronic revolution is far from having run its course.

- A computer business is becoming a chemical business.

- The number of components per chip can be expected to grow dramatically for at least ten to fifteen years.

Table 6.2
UNIT COSTS--THE PRESENT

	1975	1981
Satellite transponder	$64,000 per month/channel	$9,360 per month/channel
	1978	1980
Satellite terminal dish	$25,000	$5,000
	1960	1982
VCR	$100,000	$500
	1970	1981
Packet switching	$30.00 per communication	$3.50 per communication
	1975	1981
Fiber optics	$10.00 per cable meter	$1.75 per cable meter

Sources: P. Abelson, and A. Hammond, eds., Electronics: The Continuing Revolution (Washington, D.C.: American Association for the Advancement of Science, 1977); B. I. Edelson and L. Pollack, Satellite Communications, p. 49; L. Jordan, "Teleconferencing, Friend or Foe," Lodging (January 1982): 39-40 (175,000 to 25,000 for Earth Terminal); E. Williams, "Space Shuttle: Key to a New Industrial Era," Financial Times, March 5, 1981, p. 9; and P. Hirsch "Multimegabit Transmission User Designed Not Predicted," Computerworld, November 19, 1979, p. 10 See also "The Surge in Sony," Financial World 149, October 1, 1980, p. 43. "The unit cost must be much cheaper--about $500--and the machine should be much smaller--small enough so it can be put in a TV set."

- Electron beam writing a silicon chip may produce densities up to ten to one hundred times greater than today.

- The cost of electronics is approaching zero.

- The size of the electronics is decreasing to limits which are set by the granularity of matter itself.

With customary humility, the microelectronics industry boasts "you ain't seen nothing yet."[14]

Revenue constitutes the other side of the equation. Table 6.3 suggests that the revenue growth of a range of products and services will experience double digit rates in the 1980s. Obviously, these projections are at best educated guesses, but the prospect of revenue growth—even if the projections are half right—merits the attention of many industries. And if one adds the growth prospect of global markets to that of domestic projections ($100 billion in 1980 to treble by 1991), revenue prospects become compellingly attractive.[15]

Driven by falling costs and prospective revenues, the entry process seeks exciting returns on risk capital. It is that prospect, that profit expectation, that feeds the process over the long run (Figure 6.2).

THE MECHANICS OF MARKET ENTRY

As technology multiplies, options expand, and costs decline, firms engage in new products, terminals, networks, and services. A firm may couple different technologies to achieve product or service differentiation; this segmentation process stretches and elongates the information matrix. Technology penetrates the cost and demand of research, design, manufacturing, administration, finance,

Table 6.3
MARKET SIZE (BILLIONS)

Item	1980	Projected	
Broadcast	10.3	23.0	(1990)
TV sets, video disc, VCR	5.1	16.4	(1985)
Cable TV	2.3	21.5	(1989)
Pay TV	1.6	12.7	(1987)
Cable TV revenues	0.05	1.5	(1985)
Direct broadcast satellite	0	1.0	(1985)
Satellite services	0.20	2.5	(1985)
Teleconferencing	0.55	5.0	(1987)
Personal computers	0.75	10.0	(1988)
Home information systems	1.5	5.0	(1985)
Electronic mail	1.0	4.7	(1987)
Microcomputers	4.5	18.5	(1987)
Cellular radio	0	2.7	(1987)
Private satellite services	0.146	2.9	(1991)
Office information systems	11.3 (1983)	36.6	(1988)

Source: M. Koughan, "The State of the Revolution 1982," Channels (December–January 1982); and "Private Satellite Services," Communication News (January 1982).

and the ultimate delivery of product and services to the end user—the firm or consumer (Table 6.4).

BOUNDARY LINE EFFECT

In the past, geography and separate markets, products, and services differented firms and business organizations. Delineation between public and private sectors of the economy was readily acknowledged. Industry boundaries may have been

Figure 6.2
Market Entry

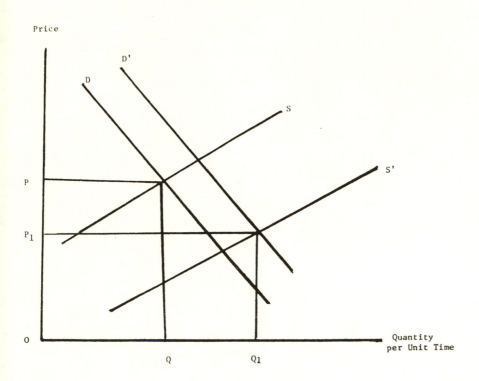

Table 6.4
MARKET SEGMENTATION

Markets	Terminals/Networks/Services
Research	Satellite scanning On-line data base
Design	Computer-aided design
Software	On-line software service
Manufacturing	Computer-aided manufacturing Local area networks Robotics, computer conferencing
Administration	Distributed data processing Electronic mail Voice mail, videotext, teleconferencing, paging
Distribution	Videotext/view data On-line inquiry, point of sale, inventory control

Finance	Point of sale, electronic funds
	    transfer, automatic teller
	    terminals (ATMs)

Home	Video text/teletext Voice/data/fax, electronic sensing

an artifact, but their existence constituted a reality that conditioned a firm's response to rival and customer alike.

INDUSTRY

Entry erodes and alters past structural boundaries (Table 6.5). We are witnessing not merely firm entry into information technology or the entry of two or three industries, rather, at least two dozen industries are positioning themselves in what might be called a multi-industry entry process.

Teleconferencing, telecommunications, and financial services highlight the multi-industry effect. (See appendixes at the end of this chapter for examples.) Consider first the diversity of firms participating in teleconferencing services (Table 6.6). Participants include IBM, Western Union, American Satellite, Holiday Inn, Marriott Hotels, Intercontinental Hotels, and AT&T.[16] Note the dispersion of industry candidates—hotel, broadcasting, computer, hardware suppliers,

Table 6.5
INDUSTRY CANDIDATES

Telephone	Retail stores
Telegraph	News services
Computer	Semiconductor firms
Minicomputer	Consumer electronics
Software	Office automation
Value added carriers	Factory automation
Radio common carriers	Hotel chains
Terminal equipment	Newspapers/magazines
Satellite carriers	Public television
Banks	Post office
Brokerage	Public radio
Insurance	Electric utilities
Cable TV	Conglomerates
Broadcasting	Airlines
Grocery chains	Data processing

telegraph carriers, minicomputer manufacturers, and even airlines indirectly through hotel subsidiaries.[17]

Second, telecommunication services reveal a common pattern in industry participation, local area networks, distribution loops, exchange systems, and local or long-haul switching facilities (Table 6.7). Firms offering these services range from Wang Computer to Xerox, from AT&T to International Harvester, from Niagara Mohawk Power to Merrill Lynch.[18] Today, some 180 firms alone offer local area network protocols to business and industrial users, and over 600 firms offer long distance communication service.[19] Industry participants range from a supplier of farm machinery to a brokerage firm, from a computer supplier to a railroad carrier, from a supplier of office copiers to the cable TV industry.

Third, financial services suggests a similar industry/firm mix—entry by a furniture store, telephone company, credit union, commercial bank, mutual fund, and a money market fund (Table 6.8).[20] Indeed, outside firms often pace telecommunications technology. In communication satellites, for example, a Citicorp engineer commented recently, "We're all amazed that to get to the state of the art we had to leave a common carrier and come to a bank."[21] And industry representation is equally widespread: computers, telephone, steel, retailing, grocery, banking, pianos, securities, savings and loan, money market funds, and insurance—a vectoring of formerly distant and unrelated activities.[22]

Spatial Boundaries

Terminals, networks, and services also erode geographic boundaries. On-line terminals access information on networks blind to spatial accident. Table 6.9 illustrates this phenomenon.

A bank may be licensed to accept deposits and engage in loans within a state;

Table 6.6
TELECONFERENCING

Firm	Industry
IBM	Computer
AT&T	Telephone
American Satellite	Satellite
Western Union	Telegraph
Video Com Net	Video broker
MACOM	Hardware supplier
Holiday Inn	Hotel
Hilton International Hotel	Hotel (TWA)
WETA TV	Public broadcasting
Allstate Insurance	Insurance
Datapoint	Minicomputer
Bell and Howell	Cameras
California Microwave	Equipment supplier
Intercontinental Hotel (formerly Pan Am subsidiary)	Hotel

but Merrill Lynch offers a nationwide financial service.[23] A savings and loan serves a confined geographic area, but an American Express terminal offers access to national and international customers. A cable TV firm may prize its local area franchise, but satellite dishes, satellite master antenna systems (SMATV), to say nothing of pay TV, and prospectively direct broadcast satellites enable firms to tap an entire U.S. market.

Although newspapers are often targeted to local readership and advertising revenues, terminals, networks, and distribution enable the print media to reach national readers—witness the national, even international, trend of the *New York Times, Wall Street Journal,* Gannett's *USA Today*, and the *Washington Post*.[24] A dry goods store often perceives its clients as residing within a 30-mile radius. Yet the growth of on-line terminals and 800 in WATS lines permits Sears Roebuck, L. L. Bean, and American Express to reach for a national retailing

Table 6.7
TELECOMMUNICATIONS

Firm	Industry
AT&T	Telephone
Pacific Power and Electric	Electric power
Non-Bell Telephone companies	Telephone
IBM	Computer
Wang Computer	Minicomputer
American Satellite	Satellite
GTE	Telephone
MCI	Specialized carrier
Internationl Harvester	Farm machinery
Merrill Lynch	Finance
Xerox	Copiers
Niagara Mohawk Electric	Power
Federal Express	Air freight
Tymeshare	Data processing
Dow-Jones	Newspaper
Southern Pacific	Railroad
COMSAT	Satellite
Manhattan Cable	Cable TV
CBS	Broadcasting
Datapoint	Minicomputers
Rolm	Telephone PBX
Northern Telecom	Telephone manufacturing
Citicorp	Bank
Sears	Retail

A local telephone company may regard its franchise as exclusive and its service market beyond penetration. Today telephone companies invade each other's markets for equipment, local area networks, local distribution facilities, and long-haul transmission services.[26]

Finally, a New York hotel may serve as a national convention site. But the teleconferencing network of Holiday Inn can tap a national audience via each of its franchised locations.[27] Satellite networks "see" the entire geography of the United States, superseding local, metropolitan, and regional markets premised upon geographic integrity and market exclusivity.[28]

Sectoral Distinctions

An information matrix blurs previously acknowledged distinctions between public and private sectors of the economy (Table 6.10). Granted that our economy has tolerated a plurality of sector institutions, market entry can unexpectedly convert public institutions into direct rivals with their private counterpart and vice versa.

WETA-TV, a PBS affiliate, offers teleconferencing, an alternative to physical

Table 6.8
FINANCIAL SERVICES

Firm	Industry
Security Pacific	Bank
Merrill Lynch	Securities
Citibank	Bank
Sears	Retail
Baldwin Piano	Piano
National Steel	Steel
Kroger's	Grocery
N. H. Credit Union	Credit union
American Express	Credit card
Prudential	Insurance
Fidelity Group	Mutual funds
Travelers Insurance	Insurance
Exxon	Petroleum
American Can	Packaging
GE	Electrical equipment
Penney's	Retail

travel. This station employs its facilities for such services on a commercial basis. Today a public institution finds itself competing with AT&T, IBM, the Hilton Hotel chain, and Marriott Inns.

IBM, GTE, and Tymeshare offer an electronic mail service to subscribers throughout the United States, but the U.S. Post Office's electronic computer originated mail (ECOM) service offers an electronic mail service that is competitive with that of the private sector.[29] Again, private/public sector institutions find themselves soliciting identical customers.

There are some 6,000 data processing firms in the United States, none of which enjoys a commanding market share. Unintimidated, National Public Radio

Table 6.9
SPATIAL EROSION

Local bank	ATM (automatic teller machine) Telephone TV set
Local savings and loan	Merrill Lynch
Local cable	Satellite
Local radio	Satellite
Local newspaper	USA Today, Wall Street Journal
Local store	Catalogue store--L.L. Bean
Local telephone company	United Telecommunications
Local hotel	Satellite Business System

has announced an information distribution service employing FM subchannel broadcast techniques and exploiting its satellite network linking over 200 stations.[30]

Whether facsimile, education, electronic funds transfer, electronic mail, satellites, or transponder sales, one pattern is discernible. Firms in the private sector increasingly find themselves competing with public sector institutions.

Global Networks

A fourth dimension of boundary line erosion is extraterritorial. Terminals, networks, and services gird our continents. Global networks permit computer-aided manufacturing, computer-aided design, office automation, on-line services transactions, satellite scanning, bibliographic data inquiry systems, airline reservations, international banking, and global financial services. That technology has dispersed knowledge is now taken for granted. The exporting and importing of software via satellite are part of that general trend.

Today, automobile manufacturers design products via international networks, information data bases are accessed worldwide, bibliographic data and processing networks abound, and financial markets in Europe, the Far East, and North America are interdependent.[31] To the extent that information networks broaden market participation, a domestic firm no longer stands isolated from global competition.

In sum, an information infrastructure and its entry process erode one's notion of boundary integrity. Industry demarcations are no longer static, spatial lines no longer isolated, and sectoral demarcations no longer beyond reach. This

Table 6.10
SECTOR EROSION

Teleconferencing	IBM, GTE, AT&T, WNET-TV
Electronic mail	IBM, GTE, Tymeshare, USPS (U.S. Postal Services)
Mail delivery	UPS, USPS, American Express
Information distribution	GE, Mead Paper, <u>Time</u> Magazine, NPR (National Public Radio)
Telephone local distribution	New England Telephone New York Port Authority
Mail boxes	Private boxes, USPS
Facsimile	Graphic scanning, USPS
Education	University of Massachusetts, Wan Institute of Graduate Studies
Electronic funds transfer	Citibank, USPS, AT&T
National paging networks	NPR, Radiofone
Satellite transponder sale	Western Union, Public Broadcasti Service, Satellite Business Systems
Satellite launching	NASA, Space Services Inc. of America

phenomenon of boundary line erosion suggests that the environment within which today's firm operates is and will experience profound alteration.

NOTES

1. "IBM in 2nd Shot at Competitors in DP Network," *Electronic News*, October 20, 1980, p. 20.

2. K. Kobayshi, "Telecommunications in the Future," *IEEE Communications* (July 1980): 103.

3. Chace, "Rising Rivalry, AT&T and IBM Tread on Each Others' Toes as Courses Converge," *Wall Street Journal,* September 4, 1981, p. 1. "The guts of modern telephones and computers are so similar says Bell Laboratories Research Director Vyssotsky, it's difficult even for a specialist to distinguish between them. 'If I know how to build a word processor,' he said, 'I know how to build a switching system.'"

4. R. T. Lund, "Microprocessor and Productivity: Cashing in on Our Chips," *Technology Review* (January 1981): 43; Jack Robertson, "Says Pentagon controls hamper VHSIC Research," *Electronic News,* August 27, 1982, p. 16.

5. R. Emmett, "Return of the Vikings," *Datamation* (April 1981): 83; L. M. Ericsson, "Di-

gesting the Microchip," *The Economist* (January 1982): 68; "Ericsson moves on the U.S. Giants," *Business Week*, March 22, 1982, p. 74; F. Baur, "Microelectronics in International Competition," *Siemens Review*, No. 6 (1980): 9; C. Lehner, "Technology Duel: Japan Strives to Move from Fine Imitations to Its Own Inventions," *Wall Street Journal*, December 1, 1981, p. 1; and "Pitfalls in France's Vast R&D Plan," *Business Week*, November 23, 1981, p. 94.

6. "Telecommunications Success Story," *Financial World*, November 15, 1981, p. 23.

7. "Com Sat Buys IC Firm," *Electronics*, vol. 55, February 10, 1982, p. 63.

8. "Rochester Telephone Expects Profit Gain in Six Months and Year," *Wall Street Journal*, March 17, 1982, p. 42.

9. "Tymeshare Acquires Microband," *Wall Street Journal*, January 7, 1981, p. 33.

10. P. Hirsch, "Major Industry Battle Brewing Over Wideband Dems Planned by Tymnet, SBS and Isacomm," *Computerworld*, September 7, 1981, p. 20.

11. F. Rose and S. Chace, "IBM Agrees to Work with Mitel to Develop Phone Equipment for Entry in U.S. Market," *Wall Street Journal*, July 22, 1982, p. 4.

12. Guy de Jonquieres, "We'll Talk to Anyone Who Has a Better Mousetrap," *Financial Times*, January 24, 1983, p. 12; and Paul Betts, "Baby Bell Starts Work on a New Image," *Financial Times*, January 20, 1983, p. 7.

13. W. G. Hutchinson, "Computer Technology to 1990," *Computer Data* (May 1979): 24.

14. John S. Mayo, "The Power of Microelectronics," *Technology Review* (January 1981): 46; and P. Laurie, *The Micro Revolution* (London: Futura Publications, 1980), p. 80.

15. F. G. Withington, "The Role of Japan in the Future World Information Processing Industry," *Report*, Arthur D. Little, Inc., December 20, 1982, p. 10.

16. *Satellite Communications* (June 1982): 27; and L. Jordan, "Teleconferencing: Friend or Foe," *Lodging* (January 1982): 37.

17. S. Cooney, "Lowering Skies for the Satellite Business," *Fortune*, December 13, 1982, p. 148.

18. "Harvester Cleared for Phone Service," *New York Times*, April 10, 1982, p. 30. "Wang to Acquire Stake in Sat. Firm," *Electronic News*, November 22, 1982, p. 12; and J. Thompson, "Fiber Optics for Power Plant Teleprotections, *Communication*," International Fiber Optics and Communications (IFOC) (September 1982): 35.

19. "Telephone's New Era: How Users Will Fare," *U.S. News and World Report*, March 14, 1983, p. 72.

20. C. Anders and M. Dodush, "Baldwin-United, in Reversal, Plans to Use Bank Credit in Financing MGIC Take-over," *Wall Street Journal*, February 9, 1982. See also P. Gigot and T. Lueck, "Sears Expansion Brings Increased Competition to Bankers and Brokers," *Wall Street Journal*, October 12, 1981, p. 1; R. A. Bennett, "Wriston Ponders a Bankless Citicorp," *New York Times*, July 31, 1981, p. D1; "Alliance of Big Banks Plans National Teller Machine Networks," *Boston Globe*, April 8, 1982, p. 27; Bruce Hoard, "ATM Networks Moving Nationwide," *Computerworld*, November 6, 1982, p. 61; T. Schell Hardt, "Thrift Cleared to Enter Securities Field by Bank Board; Legal Challenges Likely," *Wall Street Journal*, April 7, 1982, p. 21; and "The New Sears," *Business Week*, November 16, 1981, p. 140.

21. Robert A. Bennett, "Citicorp's Satellite Challenge," *New York Times*, March 24, 1983, p. D1.

22. Kentucky Food Marts Chosen to Provide EFT Services," *Management Information Systems Week*, February 3, 1982, p. 31.

23. A. Cane, "Why IBM Will Shake the Videotext Game," *Financial Times*, November 2, 1982, p. 14.

24. "Gannett's National Gamble," *Newsweek*, September 20, 1982, p. 101. See also "Wall Street Journal Plans European Edition," *Editor and Publisher*, May 1, 1982, p. 62; and "New York Times Plans to Launch Satellites at California Plant," *Boston Globe*, April 7, 1982, p. 32.

25. "Catalogue Cornucopia," *Time*, November 8, 1982, p. 71.

26. Jean A. Briggs, "The Pace Setter," *Forbes*, January 3, 1983, p. 42.

27. "Bell Asks Teleconferencing Service OK," *Electronic News,* March 23, 1981, p. 20; J. Marks, "Com Sat, Intercontinental Hotels to Offer International Teleconferencing," *Satellite Communications* (June 1982): 27; C. Jordan, "Teleconferencing, Friend or Foe?," *Lodging* (January 1982): 37; W. Arnold and J. Rausch, "Teleconferencing vs. $21 B Annual U.S. Business Travel," *Electronic Business* (June 1981): 76.

28. L. Bergreen, "Hello LA? Miami? Hong Kong?," *New York Times,* August 30, 1981, p. F5.

29. R. Wiley and D. Adams, "Should FCC or USPS Control Electronic Mail?," *Legal Times of Washington,* April 16, 1979, p. 12.

30. W. Tucker, "Public Radio Comes to Market," *Fortune,* October 18, 1982, p. 208. "NPR Set to Sell Satellite Time," *Broadcasting,* 102, January 11, 1982, p. 75.

31. A. Pollack, "Latest Technology May Spawn the Electronic Sweatshop," *New York Times,* September 3, 1982, p. 14; and "Where Future Jobs Will Be," *World Press Review* (March 1981): 23.

Appendix 1
Electronic Mail / Word Processing

Firm	Industry
BBN	Software
IBM	Computer
DEC	Minicomputer
Wang	Minicomputer
Telenet	Common carrier
Texas Instruments	IC supplier
Bekins	Moving
Federal Express	Package delivery
Tymeshare	EDP process
Datatron	Software
Xerox	Copier
NBI	Remote terminals
AXXA	City Corp (bank)
British-Leyland	Auto manufacturing

Appendix 2
Videotext/Teletext

Firm	Industry
CBS	Cable
AT&T	Telephone
New York Times	Newspaper
Time	Magazine
Northern Telecom	Equipment supplier
Sony	TV set manufacturer
IBM	Computer
Storer	Broadcast
Radio Shack	Distributor
Warner/Amex	Broadcast/credit card

Appendix 3
Computer Retailing

Firm	Industry
Radio Shack	Distribution
Computerland	De Novo
Digital Equipment	Minicomputer
Xerox	Copier
Sears	Retail
Texas Instruments	IC supplier
IBM	Mainframe
AT&T	Telephone
Hewlett Packard	Minicomputer

Appendix 4
Telephone PBX

Firm	Industry
AT&T	Telephone
GTE	Telephone
Rockwell	Aerospace
Mitel	De Novo
Datapoint	Minicomputer
Northern Electric	Telephone equipment
IBM	Computer
Rolm	De Novo
Anaconda-Ericsson	Copper supplier
Lexar	City Corp
Harris	Printing

Appendix 5
Local Area Networks

Firm	Industry
IBM	Mainframe
Datapoint	Minicomputer
AT&T	Telephone
Digital Equipment	Minicomputer
Ungerman Bass	De Novo
Xerox	Copier

Appendix 6
Software

Firm	Industry
Informatics	De Novo
IBM	Computer
Digital Equipment	Minicomputer
Tymeshare	Time-sharing
AT&T	Telephone
Disney	Movies/entertainment
CBS	Cable TV
20th Century Fox	Movies

Appendix 7
On-Line Service

Firm	Industry
IBM	Service bureau
McDonald-Douglas	Data processing
Lockheed Dialog	Aerospace
New York Times	Newspaper
United Telecom	Telephone
Mead Data	Paper
General Electric	Equipment manufacturer
Tandy	Retailer
Rochester Telephone	Telephone
Chase Econometrics	Bank
DRI/McGraw-Hill	Book magazine
The Source	Reader's Digest
Compuserve	H&R Block--tax service
AT&T	Telephone

7

THE EMERGING ENVIRONMENT

In transforming the environment of the firm, the entry process tends to destabilize the relationship of the firm to its rivals, customers, and suppliers. What then is the process of boundary erosion, how does it affect the firm's environment, and in what manner does the firm react or adjust? These issues flow out of an intensification of rivalry and competition in what might be termed the emerging environment.

THE EROSION PROCESS

Figure 7.1, divided into four quadrants, depicts the unraveling of a firm's relationship to its rivals, customers, and suppliers. Quadrant A represents a quiescent state. The environment of the firm is relatively stable, its boundary lines fixed, and its relationship to customer and supplier predictable. All participants acknowledge the market characteristics of computers, print, broadcasting, software, mail, voice, or office equipment. The environment of Quadrant A tends to be turmoil free and relatively calm.

Quadrant B suggests that the industry pot is now placed on the stove of technology. Technology's kinetic energy begins to soften past distinctions, although overall demarcations persist. The firm's relationship to its clients, suppliers, and rivals remains predictable, although something is astir.

Quadrant C reflects increased agitation. Not only are industry boundaries falling, but also the customer/firm/supplier relationship is becoming destabilized further. Suppliers may drift from the status of subcontractor to that of potential rival. The firm itself may move laterally into an adjacent industry, emerging as a direct rival to a new set of firms. Similarly, adjacent enterprises may move laterally into another firm's industry.

Quadrant D reflects a turbulent pot. The relationship of a firm to its clients and suppliers is in a state of flux. The firm is besieged by acknowledged rivals within the confines of its own industry. Indeed the firm can experience rivalry from former users and suppliers to say nothing of firms and their constituencies from adjacent or distant industries.

Figure 7.1
The Erosion of Market Relationships

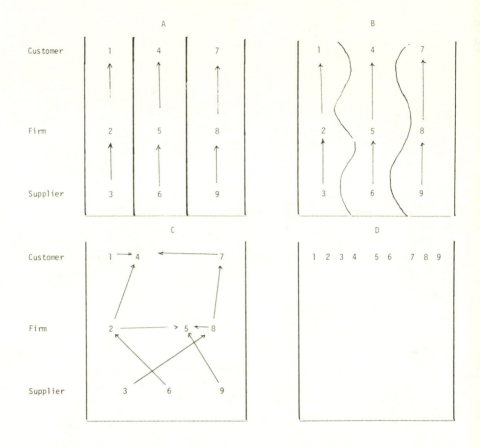

As industry lines dissolve, the process of rival multiplication gains momentum—an entry process that is reciprocal in content. A distant and presumably unrelated industry suddenly erupts as an unwelcome intruder into what a firm perceives as "its market." To the extent that the number of rivals has multiplied, a form of "blind side" competition emerges.[1] The unforeseen source of that competition only serves to augment the level of environmental risk and uncertainty. If one factors in the remaining dimension of boundary line decay—geographic, sector, global—the environmental reality of the firm has been transformed in both degree and kind.

Several illustrations point up this destabilizing process. Consider the case of a financial services firm—Merrill Lynch—as a transformation from customer to competitor. Merrill Lynch leases local and toll telephone facilities throughout the country. The company estimates that its $200 million telecommunications expense will top $1 billion in 1990.[2] In order to reduce this expenditure, the financial firm has combined with the New York Port Authority and Western Union to construct a teleport on Staten Island (Figure 7.2). The venture contemplates a set of satellite terminals that will network domestic and international sites. The teleport will link New York City, New Jersey, and other large cities via fiber optics cable, bypassing local telephone company facilities.

Given this matrix of terminals, communications, and services, Merrill Lynch will move from communication customer to in-house supplier of communications. Indeed, it is conceivable that Merrill Lynch will do more than confine its delivery system internally. To the extent that the company experiences excess capacity in both loop and long-haul satellite facilities, Merrill Lynch can emerge as a local loop reseller competing with the terrestrial facilities of New York Telephone. When that occurs, Merrill Lynch's firm/customer status will change irrevocably. A customer will have surfaced as a direct rival to the telephone company.

A manufacturer may also become a competitor. Texas Instruments supplies integrated circuits to both IBM and AT&T products. Over time, Texas Instruments has evolved into a manufacturer of consumer end products, specifically personal computers.[3] To the extent that the company competes with IBM in the production of such terminals, the firm/supplier relationship has been transformed and altered.

Or consider two telephone companies, each offering services into two distinct geographical areas. In the past, geographic separation by definition removed any perception of direct competition, but the movement of each into local access areas, satellites, data processing, and subscriber equipment places each on a collision course.

But carry the entry process further to unrelated firms in once distant and far removed industries. Corning Glass traditionally supplied silicon products to the television industry and research laboratories. Its work in fiber optics has prompted the firm to manufacture and market optical cable throughout the world. The company has evolved into direct competition with copper cable manufacturers and communications satellite suppliers.

The case of Exxon also illustrates the distant boundary phenomenon. Exxon's Gilbarco Division supplies gas pump apparatus for retail stations. Combining microprocessors, software, and networking,[4] Gilbarco is moving into an electronic funds transfer prototype that may lend itself to operating on a regional or national basis. A gas pump manufacturer is emerging as a direct rival to a bank, a telephone company, or a computer firm. Indeed—the gas pump is now an automatic teller terminal, witness NCR's system in Portland, Maine.[5]

Federal Express underscores the phenomenon of blind side competition. The company painted its airplanes purple to ferry small packages to Memphis, Tennessee, for sorting assignment and next-day delivery. Although the firm competed with the post office, it stood as an acknowledged customer of the nation's telephone companies. But Federal Express estimates that 40 percent of its revenues are derived from documents that lend themselves to electronic transmis-

Figure 7.2
Customer as Potential Competitor

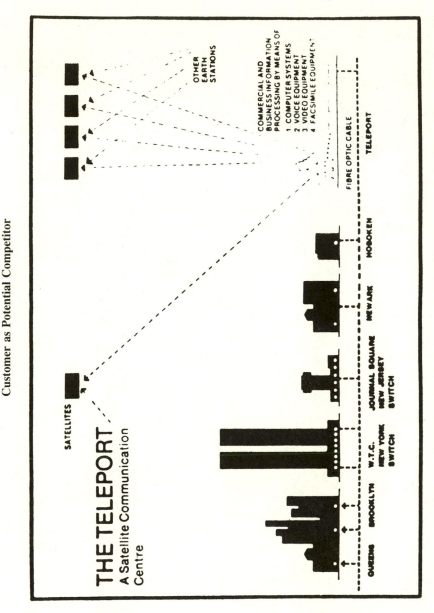

Source: Paul Betts, ''Beaming on Staten Is.'' *Financial Times*, December 8, 1982, p. VI.

sion.[6] Hence, the company is leasing satellite transponders and is applying for a digital termination system license from the FCC to bypass telephone company distribution facilities. An air courier system will undoubtedly emerge as a direct competitor to the telephone industry. Perhaps this explains why Bekins, Inc., a household mover (subject to ICC jurisdiction) is similarly diversifying into electronic mail as an extension of its physical transfer of household furnishings.[7]

The electric power industry and telephone carrier appear separate and distinct in markets and services. On the federal level, each industry is subject to separate regulatory bodies, laws, and commissions (FERC [Federal Energy Regulatory Commission] and FCC). Today, electric power companies are stringing fiber optic cables along their transmission rights of way. Fiber optic cable yields 5,000 times more capacity than comparable copper cable.[8] Undoubtedly, power utilities will experience excess capacity. Is it not conceivable that electric power grids linked by satellite could provide voice, data, record, video, and document transmission service in direct competition with cable TV and telephone companies? The fact that electric power installations are incorporating fiber optics argues that such a possibility only awaits the passage of time.

Not unlike power utilities, U.S. railroad carriers are discovering that their track rights of way are suddenly an invaluable resource. Fiber optic networks linked by satellite, cellular radio, or digital termination systems, hold the promise of new information services. The result is that railroads are moving from customers of the nation's telephone companies to the status of competitor and rival. Currently, fiber optic networks are proposed by the Missouri, Kansas, Texas, Amtrak, and Southern Pacific. Surely other railroad carriers, in following suit, will participate in an emerging information environment.

Finally, video conferencing by satellite suggests that unrelated industries can collide unexpectedly. As we have noted, the hotel industry offers video conferencing as an extension of their investment in satellite dishes, transponders, and meeting facilities. Yet the Allstate subsidiary of Sears concludes that its customers will be in "good hands" if they lease voice, data, and video services from the company. Indeed, Allstate plans to set up some 30 videoconferencing rooms, scattered throughout the United States, containing screens, monitors, and document display systems.[9] In a real sense, the hotel industry, the insurance industry, the telephone industry, and the airline industry stand as direct competitors.

But as boundaries soften, the CWA finds itself attempting to organize government employees, employees in office automation, and semiconductor personnel.[10] In so doing, the CWA is encountering competition from the United Auto Workers, International Brotherhood of Electrical Workers (IBEW), the Teamsters, and the Machinists Union. Falling boundaries increasingly destabilize union jurisdictional lines.

In sum, technology and market entry impart a profound effect on the environment of the firm. Relationships uncouple not only within a single industry but also among adjacent and even remote industries. The erosion process pro-

Table 7.1
INNOVATION CYCLES

Local Switching		Toll Switching	
Manual	1880s	Manual	1880s
Step	1919	Step	1926
Panel	1921	X-Bar Tandem	1941
No. 1 X-Bar	1938	No. 4 X-Bar	1943
No. 5 X-Bar	1948	No. 5 X-Bar	1953
No. 1 ESS	1965	No. 1 ESS	1970

Source: M. Fagan, A History of Engineering and Science in
the Bell System (Bell Telephone Laboratory, 1979), p.
54.

duces an environment of intensified, even frantic competition. That new competition marks the essence of an emerging environment.

THE EMERGING ENVIRONMENT

What is the effect of borderless markets and intensified competition upon research, price, cost, obsolescence, and marketing, as well as plant location? First, intensified competition contracts product development cycles. Lead times of ten to fifteen years, long common with the telephone company, no longer appear appropriate today (Table 7.1). In one sense, research cannot be disassociated from the innovation process.

Second, intensified competition accelerates the flow of new products to the marketplace and renders products once regarded as viable and long lasting as obsolete and write-off candidates. Product life shortens and contracts as noted in Table 7.2.

New products, competing with existing hardware, inject an element of risk into any firm's decision-making process. In an environment of discontinuity, even "progressive" firms can misstep; Texas Instruments in home computers, IBM in minicomputers, AT&T in PBXs, Northern Telecom in office automation, Fairchild Camera in digital watches, and Siemens Corporation in telephone switches.[11]

Third, intensified competition accelerates the pace of cost and price adjustments. Pressure to increase productivity, cut unit cost, reduce prices, and tap

Table 7.2
PRODUCT INNOVATION

Semiconductors	4 generations in 5 years
Mainframe computers	1 generation in 4 years
Minicomputers	1 generation in 2-3 years
Microprocessors	3 generations in 5 years
Packet switching	3 generations in 8 years
Satellite	5 generations in 20 years
Printers	1 generation in 18 months
IC test equipment	1 generation in 12 months
PBXs	1 generation in 5 years
CRT terminals	1979 36 months--1982 18 months

Source: M. R. Irwin, "U.S. Telecommunications Policy:
Technology vs. Regulation," Information Society:
Changes, Chances, Challenges, Fourteenth
International Netherlands Organization for Applied
Scientific Research Conference, Rotterdam, The
Netherlands, 1981.

new markets sometimes borders on the furious. Hand calculator prices (Figure 7.3) and the experience of digital watches, video games, and VCRs merely reflect this price/cost cycle.

Fourth, intensified competition introduces product, service, and process innovation as key management endeavors. It is one thing to commit research, development, manufacturing, and distribution decisions in a world of discretionary change. In a universe of fluid consumer preferences and expanded product options, marketing often emerges as a crucial variable. Marketing embraces a host of often contradictory traits—the intuitive, the artistic, the perceptive, and the decisive.

Innovation may originate from the customer side of the equation. A recent MIT study concludes that the customer originated 100 percent of sixty-six "major" innovations, 35 percent of sixty-six major improvements, and 60 percent of eighty-three "minor" improvements. Clearly, any firm able to anticipate

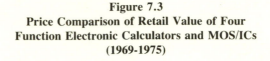

Figure 7.3
Price Comparison of Retail Value of Four
Function Electronic Calculators and MOS/ICs
(1969-1975)

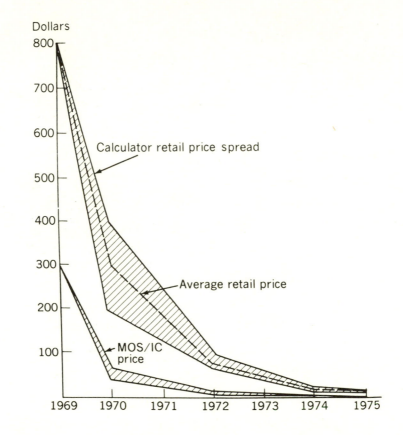

Source: Selected issues of Electronics, Electronic News, and information from industry sources.

present and future customer needs possesses an advantage in an environment of accelerating change.

Fifth, intensified competition yelds an abundance of consumer options and alternatives. As technology expands and multiplies, the range of user options grows (Table 7.3). This expansion opens avenues to the consumer, as well as the business firm.

Finally, the emerging environment is one of investment fluidity. Networks permit the collection, storage, and distribution of information that supersede geographic parochialism. Corporations can relocate as long as network accessibility is available. Knowledge intensive industries locate near knowledge intensive institutions; and the university, the science park, the research area, and

Table 7.3
USER OPTIONS

Video disc versus VCR

Travel versus teleconferencing

Home video versus university lecture

Videotext versus banking

Electronic mail versus post office

Broadcast versus cable TV

Newspaper versus teletext

Cellular radio versus telephone wire

Individual-design versus computer-aided design

Computer versus tax service

Videotext versus shopping

Home security versus personal guard

On-line bibliographic service versus public library

Direct broadcast satellites versus commercial TV

Facsimile versus letter

Local area networks versus PBX

---

the teleport can serve to determine industry and investment decision-making.[12] To that extent, investment and plant decentralization are facilitated by an emerging information infrastructure.

How then can one sum up the anatomy of this emerging environment? We suggest the following characteristics:

- A firm's product or service is likely to experience more, rather than fewer, competitive substitutes in the 1980s.
- A firm can no longer be certain of its industry boundary.
- A firm may find it essential to address sectors of the economy rather than individual markets.

- Competition will be augmented by rivals from one's customers and one's suppliers.
- Competition will erupt from the constituents of distant industries—firms, customers, and suppliers.
- Intensified competition will contract R&D cycles.
- Intensified competition will narrow production runs.
- Intensified competition places a premium upon a firm's marketing and innovation response.
- Intensified competition will enhance risk uncertainty but promises inviting returns on investment.
- Intensified competition accelerates capital mobility and industry decentralization.

RESPONSE OF THE FIRM

The response of firms to an emerging environment constitutes a study unto itself. An initial impression suggests, however, that today's corporations are reassessing their perceptions of market opportunity, diversification, and asset positioning. Industry boundary loss obviously affects virtually every investment decision of the firm, and boundary erosion prompts a reevaluation of research product development, plant location, marketing strategy, make/buy decisions, and product distribution. In an environment of dynamic change, flexibility, innovation, productivity, and entrepreneurship emerge as critical and important assets.

The Existing Firm

But how do existing institutions balance centralized control with decentralized operations? How can a firm blend creativity on one side and stability and order on the other? Not surprisingly, many firms seek to expedite the decision-making process by flattening their hierarchical structure, reducing layers of management, pushing investment and marketing decisions deeper into corporate divisions, and dividing the organization into smaller, autonomous units.

Some firms endeavor to institutionalize innovation by privately sanctioning "under desk" projects or unauthorized "skunk works." One method of diversifying risk is to pursue changes on the margin. 3M is fond of saying, "We make a little, sell a little, and make a little more."[13] Institutionalized innovation, marketing, entrepreneurship, and risk within the existing corporate structure pose elusive, but ongoing, exercises.

Joint Ventures

Still another diversification of risk is to employ the joint venture as a strategic commitment. Within the past two years, the joint venture, technical agreements, and exchanges have seen greater use and application (Table 7.4).

If nothing else, the joint venture enables an organization to secure expertise, to spread R&D cost, to share management talent, to move into untapped geo-

Table 7.4
JOINT VENTURES/JOINT AGREEMENTS/TECHNICAL EXCHANGES

Firm	Manufacturing/Terminals	Networks	Marketing/Services
IBM/Nippon	Fax terminals		
Scientific Atlanta/ Plessey		Cable/satellite	
Zenith/Taft			Teletext
Burroughs/Intel	IC chips		
Xerox/Intel/DEC		Local area network	
AT&T/CBS			Videotext
Texas Instruments/IBM		Local area network	
Microdyne/Domesticom		Master satellite	
	antenna systems		
New York Times/CBS			Video text
Western Union/Airfone			Air telephone service
Northern Telecom/GE			Cellular radio
OAK/RACAL		Pay TV/satellite	
GEC/Marconi/Nippon			Cellular radio
Ericsson/Anaconda	Telephone systems		
Timex/Sinclair	Personal computers		
American Satellite/ Continental Telephone		Satellite net	
GCA/Matra	Semiconductor equipment		
Intel/NEC	Microprocessors		
Motorola/NEC	Pocket pagers		
AT&T/GTE			Cellular radio
AT&T/N. V. Philips		Digital switch	
Sperry/Mitsubishi	Office automation		
Siemens/Western/Digital	IC chips		
Columbia/HBO/CBS			Movie software
AT&T/Plessey			Cellular radio
Com Sat/IBM/Aetna		Satellite net	
Alcatel/N. V. Philips		Cellular radio	
	switches		
IBM/Kanematsu-Gaslto			Electronic typewriter
Control Data/N. V. Philips	Recording drives		
GTE/Italtel		Central office	
	switches		
IBM/ROLM	PBXs		
American Express/Warner Communications			Videotext

Source: "IBM and Intel Link Up to Fend Off Japan," _Business Week_, January 10, 1983, p. 97; and "Giants Enter Ventures as Partners in Prosperity," _Electronic Business_ (January 1983): 74.

graphic markets, to buy time. However popular, the jury is out as to the effectiveness of this approach over the long pull. As Table 7.4 suggests, the joint venture approach is not merely a domestic but a global phenomenon as well.

An environment of intensified competition places a premium upon a firm's ability to respond, react, and adjust. In a world of moderate, discretionary change, a company might elect to incorporate development, production, and marketing in-house. But as the pace of the emerging environment quickens, a real question is whether integration adds to or subtracts from a firm's flexibility. Increasingly, firms are reassessing vertical integration and their attendant make/buy decisions.

IBM provides an insightful illustration. The company subcontracted out major components of its personal computer circuitry, hardware, software, and distribution.[14] Indeed, IBM created a separate organization to deal with the imperatives of an intensely competitive market. As one observer put it, today's firm must concentrate on market innovation. . . ."And get somebody else to do the metal bashn'."[15] Still other companies insist that one key to cost efficiency resides in the subcontracting process.[16] Whatever the strategy, the virtue and liability of corporate organizations is coming under review and scrutiny.

The New Firm

The entrepreneurial firm flourishes in this borderless environment, and the new corporation is crucial to the innovation process. Start-ups are fraught with risk, nevertheless. Only 40 percent of all new firms are said to survive the first five years. On the other hand, new firms often set the pace in the introduction of new products, service, and new processes. This is not to suggest that the incumbent firm is unimportant to the innovation process, but rather to emphasize that start-up firms perceive opportunities that going concerns often overlook.[17]

More systematic studies bear out the innovative record of venture capital and the small firm. An Office of Management and Budget study concludes that firms with under 1,000 employees account for 50 percent of the major technical innovations in the United States; an MIT study has shown that the small firm is four times as innovative per R&D dollar as a large corporation; and a National Science Foundation study has revealed that the small firm enjoys twenty-four times more innovative breakthrough than the large firm (10,000).[18]

Decisions by venture capital and prospective firms are critical to job generation as well. An MIT study observed that from 1969 to 1976 66 percent of all new employment came from firms employing less than fifty people. (Table 7.5).[19] More importantly, the MIT study concluded that 80 percent of all new jobs came from companies less than five years old.[20] Though less optimistic, a Brookings Institution study concluded that from 1976 to 1980 51 percent of new jobs were generated by firms employing less than 100 individuals.[21] A similar pattern was revealed by a Canadian study that concluded that 72 percent of all new jobs came from companies employing less than twenty individuals. The Irish Industrial Authority concluded in 1982 that 30 percent of its new jobs were derived from companies employing fifty or less individuals.[22]

Venture capital and the small firm make signal contributions to exports. Fifty percent of all Japanese exports derive from firms classified as small business; 50 percent of the United Kingdom's Queen's Award for exports went to firms with under 200 employees; and 70 percent of Japanese production workers are employed by firms with under 300 employees. (The U.S. percentage is 40 percent).[23]

An environment of dynamic change places a premium upon enterpreneurship, ingenuity, and creativity. If, as industry participants assert, technology is unlikely to slow down in the foreseeable future, companies must adjust to an environment of opportunity, change, turbulence, and uncertainty.

In sum, shifts in demand and supply, driven in large part by explosive changes in technology, feed an entry process that renders asunder product, industry, sector, or geographic demarcations. The blurring of boundaries alters the content of competition, the pace of market adjustment. Any firm, particularly a monopoly

Table 7.5
PERCENTAGE OF JOBS CREATED BY SIZE OF FIRM AND REGION

Number of Employees in Firm	Percentage of Jobs Created			
Northeast	North Central	South	West	U.S. Average
0-20 177.1	67.2	53.5	59.5	66.0
21-50 6.5	12.0	11.2	11.6	11.2
51-100 -17.3	5.2	5.5	6.3	4.3
101-500 -33.3	3.1	9.4	9.3	5.2
501+ -32.9	12.4	20.4	13.3	13.3
Total 100.0	100.0	100.0	100.0	100.0

Source: D. Birch, "Who Creates Jobs," The Public Interest

(Fall 1981): 8.

firm, can forfeit its discretionary control over product innovation, service pricing, equipment obsolescence, and market positioning.

As this reality began to penetrate the Bell System, it was clear the AT&T rivals could adopt smart missiles, scanning satellites, and microelectronics countermeasures while Bell was confined to the attrition of barbed wire, machine guns and trench warfare. As the process continued, a reassessment of the 1956 Consent Decree appeared all but inevitable.

NOTES

1. John Ela, Strategic Planning, Gilbarco, Inc., first employed the phrase "blind side competition." See J. Ela and M. Irwin, "Blindside Competition: Technological Change Demand Different Strategies in Information Age," *Marketing News*, November 26, 1982, p. 10.

2. Paul Taylor, "Beaming on Staten Is.," *Financial Times*, December 8, 1982, p. 7. See also R. A. Bennett, "Citicorp's Satellite Challenge," *New York Times*, March 24, 1983, p. D1.

3. "Texas Instruments Comes Roaring Back," *Business Week*, February 14, 1983, p. 110; Elaine Williams, "Texas Ready to Compete," *Financial Times*, February 3, 1983, p. 18; "Intel Moves Its Skills into Black Box Field," *Electronic Business* (June 1982): 106; and Gene Bylinsky, "Fiber Optics Finally Sees the Light of Day," *Fortune*, March 24, 1980, p. 111.

4. "Price Setting for First Throughput," *Financial Times*, March 11, 1983, p. 23.

5. "Gasoline Station to Use NCR Spinoff of Teller Machines," *Wall Street Journal*, May 2, 1983, p. 8.

6. "Telecommunications," *Business Week*, October 11, 1982, p. 60; and C. Stipp, "Helloooo Electronics, Federal Says," *Wall Street Journal*, April 28, 1982, p. 13.

7. "A Household Moving Leader Highballing into Related Services," *Business Week*, November 1, 1982, p. 58.

8. C. Charlish, "EGB Moves Messages with Power through Grid," *Financial Times*, September 27, 1982, p. 12; and A. Fairaizl, "Electric Utility Remote Location Interconnection with Fiberoptics," *IFOC* (September 1982): 29.

9. Kathryn Jones, "Allstate Jumps Into Voice, Data Market," *Management Information Systems Week*, June 15, 1983, p. 26.

10. S. S. King, "Labor Deals with Hard Times, New Technology," *New York Times*, February 20, 1983, p. E3; Ann Dooley, "Union Leader Sets Sights on the Automated Office," *Computerworld*, p. 8; L. Waller, "Union Activity Astir on West Coast," *Electronics*, April 7, 1982, p. 40; "Unions Move into the Office," *Business Week*, January 25, 1982, p. 90; "A Union Fight That May Explode," *Business Week*, March 16, 1981, p. 102; W. P. Patterson, "A New Worry for Chipmakers," *Industry Week*, July 27, 1981, p. 82; and "A Union Gets Ready for the Information Age," *Business Week*, July 26, 1982, p. 23.

11. Anne Bagamery, "Texas Instruments in Mid-life," *Forbes*, March 15, 1982, p. 64; Bro Uttal, "Texas Instruments Regroups," *Fortune*, August 9, 1982, p. 40; C. Lorenz, "Why ITT Finds That Being Big Is Not Enough," *Financial Times*, June 6, 1982, p. 8; and "No. 1's Awesome Strategy," *Business Week*, June 8, 1981, p. 86. See also Bro Uttal, "A Computer Gadfly's Triumph," *Fortune*, March 8, 1982, p. 74; M. Magnet, "Clive Sinclair's Little Computer That Could," *Fortune*, March 8, 1982, p. 78; "That Mad, Mod MUX World," *Datamation* (December 1981): 67; "Jogging Entrepreneur Is Now Running More for His Local Networking Firm," *Data Communications* (September 1980): 51; A. Furst, "A Fresh Crop of Fail-Safe Computers," *Electronic Business* (October 1981): 58; and "Taking on the Industry Giant," An Interview with Gene M. Amdahl, *Harvard Business Review* (March/April 1980): 82; Ronald Rosenberg, "IBM Road Show Comes to Town," *Boston Globe*, April 24, 1983, p. 77; and T. Peters and R. Waterman, "What's Right with Big Business," *The Washington Monthly* (December 1982): 45.

12. "Business and Universities: A New Partnership," *Business Week*, December 20, 1982, p. 58; and "Academia-Industry Links Lure Companies to Ulster," *Electronic Business* (October 1982): 146.

13. T. Peters and R. Waterman, "What's Right with Big Business," *The Washington Monthly* (December 1982): 41.

14. "IBM Abandons Its Go-It-Alone Stance," *Business Week*, March 14, 1983, p. 41.

15. Myron Magnet, "Clive Sinclair's Little Computer That Could," *Fortune*, March 8, 1982, p. 84.

16. Bro Uttal, "A Computer Gadfly's Triumph," *Fortune*, March 8, 1982, p. 75.

17. Starting a New Business—Pitfalls to Avoid," *U.S. News and World Report*, July 13, 1981, p. 75.

18. "Five Ways to Go Bust," *The Economist*, January 8, 1983, p. 11; Roger Collis, "Risk Takers Find Firms Favor the Conservatives," *International Herald Tribune*, October 27, 1982, p. 8; and J. Brinton, "R&D: Some Strategies That Work," *Electronics*, vol. 53, April 24, 1980, p. 94.

19. David Birch, "Who Creates Jobs," *The Public Interest*, No. 65 (Fall 1981): 3.

20. R. Collis, "A Bid to Revitalize U.K.'s Small Business Sector," *International Herald Tribune*, October 27, 1982, p. 25; and M. McAbee, "Pebbles Support Japan's Monolith," *Industry Week*, May 1, 1978, p. 40.

21. "In Jobs, Small is Beautiful," *New York Times*, March 13, 1983, p. F23; also C. Armington and M. Odle, "Small Business—How Many Jobs," *The Brookings Review* (Winter 1982): 14.

22. R. Collis, "A Bid to Revitalize U.K.'s Small Business Sector," p. 95.

23. James McDonald, "Export Success Ranging From High Technology to Throat Lozenges," *Financial Times*, April 21, 1983, p. 12.

8

AN IMPEDIMENT TO DIVERSIFICATION

Driven by technology, a borderless economy imposes a fundamental reassessment upon any firm's market strategy. For AT&T, however, corporate diversification was conditioned by its 1956 Consent Judgment with the Department of Justice. To retain its manufacturing affiliate, Bell had elected to confine itself to regulated communication services. The price of retaining vertical integration—and hence Theodore Vail's strategy—appeared eminently reasonable and farsighted at the time. But as boundaries began to wither, what appeared to be a brilliant management coup in the 1950s was showing signs of evolving into a corporate nightmare in the 1980s. A Bell spokesman was to recall later, "We aim to be in information management, voice, data, office sensors, teleconferencing. We've known where we're going since 1975."[1]

The question persisted: could the Bell System diversify within the constraint of the 1956 decree? Could due process—which since the turn of the century had effectively protected the regulated firm—be transformed into an instrument for purposes of corporate diversification? Could flexibility be grafted onto the Consent Judgment to permit movement into computer terminals, intelligent telephones, electronic funds transfer services, and on-line data base retrieval? The answer was to be found in Bell's endeavor to file tariffs on the following services:

- The teletype model 40/4
- The transaction telephone system
- The Dimension PBX
- The advanced communication system
- Electronic yellow pages.

MOD 40

The Teletype Corporation of Western Electric, acquired in the 1930s, was a major supplier of teletypewriter and teleprinter equipment. Such send and receive devices were tied to communication lines or teletypewriter circuits. The machine represented the epitome of the electromechanical era—keys, hard copy, and response units.

In the 1960s the computer industry introduced CRT terminals, soft and hard copy, and by the late 1960s, as costs fell, CRTs became ubiquitous terminals linked to host computers. Watching its market share erode, the Teletypewriter Division of Western Electric announced a new CRT terminal to be sold to AT&T's Long Lines Division. AT&T in turn filed a tariff before the FCC.[2] Dubbed the Teletypewriter Model 40, the terminal included a display and printer unit linked to communication circuits. A year later the tariff was updated to include clusters of Mod 40s.

The Mod 40/4 filing collided with the computer industry. Computer manufacturers insisted that the Mod 40 was data processing rather than a communications device. Since data processing was unregulated, AT&T, under its Consent Decree, could not offer the service as a tariffed filing. AT&T insisted that the Mod 40 was simply an evolutionary step in its data terminal business, that tariffs for teletypewriter-printer services had been filed for years, and that the Mod 40 fell legitimately under the jurisdiction of the Communication Act of 1934.[3]

The FCC's Common Carrier Bureau took exception to AT&T's definition of the Mod 40. On the basis of their software content, the Common Carrier Bureau classified terminals as "smart" and "dumb." A terminal without software was dumb; a terminal imbedded with software was smart. Bell, the Common Carrier Bureau maintained, could file a tariff on dumb terminals only. A smart terminal constituted data processing, and under AT&T's Consent Decree, it was ruled off limits.[4]

Between AT&T's filing of the Mod 40 and a full commission ruling, Bell operating companies filed Mod 40 tariffs before thirty-nine state commissions within a six-week period.[5] State PUCs accepted the tariff as a legitimate communications service proposal; and by the time the full FCC was ready to address the issue, the commission found itself confronting state jurisdictions that had ruled the device regulated "communications."

The commission accepted the Mod 40 tariff as a communications offering. Over the objection of IBM and the computer industry, the FCC expanded its regulation to include Bell's computer terminal.[6] The FCC was obviously uncomfortable with the migration of digital and memory logic from computer to terminal but, nevertheless, held that the Mod 40 device represented a regulated activity, subject to further investigation.[7]

TRANSACTION NETWORK SERVICE

In the mid-1970s AT&T introduced its Transaction Network Service (TNS) which was oriented to banking automation. This check verification service included terminals, a host computer, and communication lines as a package offering. The service also included software protocol, conversion, and remote diagnostics. Product terminals could access a host computer, or the host computer could poll terminals for data and credit information as part of an electronic funds transfer system. The telephone company filed TNS tariffs in Washington and Minnesota; the tariffs were accepted on the premise that TNS was a regulated communication carrier appropriate to the telephone industry.

Once again, the computer industry opposed Bell's offering, insisting that TNS constituted data processing, that data processing was unregulated, and that TNS resided beyond the reach of AT&T's 1956 Consent Decree.

When an interim report from a congressional Electronic Funds Transfer (EFTS) commission examined the issue in 1977, it ruled that Bell should establish a separate subsidiary via congressional legislation.[8] Computer equipment suppliers had insisted that credit verification equipment fell within the data processing rather than communications category. Later, however, the full committee amended its position, insisting that if the FCC accepted a tariff on TNS, then Bell System companies could provide these activities to the banking industry. In effect, the EFTS Commission deferred to the Federal Communications Commission the question of whether a credit on-line verification system constituted communications or data processing. It was also clear that the Consent Decree could prove a stumbling block to Bell's ability to diversify into newly emerging markets.

DIMENSION PBX 2000

The upgrading of a telephone switch further emphasized that the 1956 Consent Decree was a mixed blessing. AT&T had offered PBX equipment to its business customers since the turn of the century. In 1977, New York Bell Telephone Company filed a tariff on the Dimension PBX 2000, a switch oriented toward the hotel market.

The Dimension tariff embodied a computer software package that permitted data collection, energy conservation, long distance toll accounting, maid location, room service and status, and other processing features. Thus, the package embraced logic and memory which permitted storage and processing capabilities as well as conventional switching features.

IBM petitioned the FCC to assume jurisdiction over the state filing of the Dimension PBX. The computer company argued that the software package was essentially electronic data processing and was beyond the definition of "communications."[9] The computer industry in fact began to ask that the AT&T Consent Judgment be lifted so as to permit Bell to diversify into new services outside of public utility regulation. The computer industry was apprehensive that Bell's diversification expanded regulation and eventually would envelop competitive firms with regulatory oversight.

ADVANCED COMMUNICATION SYSTEMS

The proliferation of terminals, the diversity of manufacturers and suppliers, and on-line devices created what some termed the Tower of Babel Syndrome. Computers, terminals, and peripheral devices could not access each other because of incompatible speeds, codes, protocols, and character recognition. Each terminal spoke a different language.

Bell's Advanced Communication Systems (ACS), originally the Bell Data Network, was introduced as a means to solve the communication disparity between different computer manufacturers. AT&T announced software and protocol conversion for purposes of compatibility and asked the FCC for a declaratory ruling as to whether the service fell within the Communication Act of 1934. AT&T reminded the commission that the FCC held jurisdiction over "communication and processing" and that processing included data processing, and AT&T advised the commission to adopt a broad rather than narrow definition of communications.[10]

Intervenors opposed Bell on the grounds that ACS constituted data processing, not communications, and resided beyond AT&T's presumed ability to diversify or participate. Bell withdrew its ACS filing with the announcement that software complexity had forced some delay. (Subsequently, the FCC in its Computer Inquiry II lifted the Bell Consent Decree and permitted AT&T to offer the service through the creation of a separate subsidiary.)[11]

ELECTRONIC YELLOW PAGES

TV sets tied to computers via communication circuits constitute an international as well as a U.S. development. Viewdata, or in the case of broadcast transmission, teletext, permits information via communication to reach the home, office, factory, or store. Home shopping, mail, security, banking, pose as an inviting market.

In 1980, Southwestern Telephone and Telegraph, a Bell operating company in Texas, filed an experimental tariff for the delivery of computerized information to selected homes in Austin, Texas. The tariff was originally accepted by the state public utility commission.

Several intervenors, Datapoint, Radio Shack, and the Texas Newspaper Publishers Association, objected that Southwestern Bell's Viewdata constituted an infringement on their future markets.[12] Each intervenor insisted that no regulated firm should combine and package information transmission with information content. The state PUC, apparently convinced by these arguments, backed off from accepting the tariff, and Bell withdrew its experimental offering from the state of Texas. The Bell operating company insisted that the PUC had rendered it "irreparable harm."[13] Once again, the Consent Decree was at issue. Did the decree bar AT&T from diversifying into on-line information services and equipment? The electronic yellow pages experiment suggested that the 1956 judgment might prove an impediment even at the state regulatory level.

In these several instances, AT&T's attempt to diversify into new markets, products, and services repeatedly encountered opposition. The burden of proof always seemed to rest with the telephone company. Whether in CRTs, PBXs, credit information, videotext, data communications, or computer protocol systems, the Consent Decree began to act as an impediment to diversification rather than as a management coup that had immunized AT&T from competitive assault.

Bell's initial strategy was to work within the constraints of the 1956 judgment. AT&T defined its new services by labeling them common carrier communications or by reminding the commission of its broad jurisdiction over "communication processing." Semantics served to confer diversity and flexibility on the firm and appeared acceptable at the state PUC level. But diversification was challenged repeatedly at the federal level; and due process was now becoming a time-consuming effort. Hearings, briefs, exhibits, and cost data, all subject to judicial review, cast uncertainty as to the legitimacy of a proposed service. Further, due process consumed time.

The premise of regulation was beginning to move from a corporate asset to a corporate liability. True, Bell had employed due process to erect economic disincentives to market entry in the past. But now due process was emerging as an impediment to AT&T's own future. A company spokesman later put it: "We cannot live with that Decree (1956). Cannot. The technology of telecommunications has so merged with the technology of data processing that if we end up with the 1956 Consent Decree we are a withering corporation waiting for its demise and nothing more."[14]

Would the Consent Decree foreclose Bell from participating in business information markets, which by the mid-1980s were estimated at a third of $1 trillion? Would the Consent Decree pose as an emerging barrier to new information services? And would the task of educating PUCs, state or federal, be so time-consuming that any initial competitive advantage would be dissipated by regulatory due process? Whatever the answer, conditions were becoming ripe for the settlement of a nagging antitrust trial and the removal of a 1956 legal bind.

NOTES

1. Marilyn Harris, "McGill Charts American Bell's Strategy," *Electronics*, April 21, 1983.

2. FCC, In the Matter of AT&T Revisions to Tariff FCC Nos. 260 and 267, *Memorandum, Opinion and Order*, March 3, 1976, p. 7; in 62 FCC 21, 1977, hereafter cited as FCC mod 40/4 Decision. "The Dataspeed 40/4 is thus an offering of data processing services." See R. Frank, "IBM, CBEMA Tell FCC not to Tariff Bell Dataspeed 40," *Computerworld*, January 12, 1976, p. 23.

3. FCC, In the Matter of Amendment of Section 64.702 of the commission's Rules and Regulations (Second Computer Inquiry) Docket 20828; Comments of AT&T, June 6, 1977, pp. 117–126. (Hereafter cited as FCC Computer II.) "Common Carriers may engage in the furnishing of processing including data processing if providing communications services within the meaning of the Communications Act."

4. A. Perlman, "Dataspeed 40/4—DP or not DP—Appeals to Panel to Decide," *Electronic News*, November 12, 1975, p. 30.

5. Vivienne Killingsworth, "Corporate Star Wars: AT&T vs. IBM," *Atlantic* (May 1979): 72. See also John Eger, "Bell's End Run," *Datamation* (May 1977): 81.

6. FCC mod 40/4 Decision, p. 30. "The data processing functions as utilized in the Dataspeed 40/4 device unquestionably results in a more efficient utilization of the connecting channel which

links the remote access device to the computer terminal. Nevertheless a data processing service is not being offered."

7. IBM vs. FCC No. 77-4405, Second Circuit, January 4, 1977.

8. George M. Dick, "TNS Developed for Short Inquiry Applications," *Computerworld*, November 28, 1977, p. S113. EFT in the United States, The Final Report of the National Commission on Electronic Funds Transfer, October 28, 1977, Washington, D.C., p. 166 (hereafter cited as EFTS Final Report). See also National Commission on Electronic Funds Transfer, Suppliers Committee Public Hearings, December 14, 1976, San Francisco. "TNS would meet FCC regulations if offered as an interstate service," p. 166. See also *Telecommunications Reports* 1943, No. 38 (September 26, 1977): 41: "The National Commission on Electronic Funds Transfer persuaded perhaps by arguments from the AT&T and the Department of Justice has decided to recommend letting AT&T participate in the EFT market essentially under the current ground rules."

9. Comments of Telenet Communications Corporation, National Commission on Electronic Funds Transfer, December 14–16, 1976, p. 15. See also testimony of the Addressograph-Multigraph Corporation to the National Commission on Electronic Funds Transfer, Suppliers Committee, "Competition Among Vendors, Request for Comment 1977." p. 28. "The primary obstacle to implementing this alternative—(Competition)—is the fact that AT&T has already tariffed EFT terminals and services and cannot in today's environment offer them thru an unregulated subsidiary. Congress should enact legislation which modifies the Consent Decree and the FCC rules to permit the sanction of this unregulated arm's length subsidiary," Ibid. p. 256. See also IBM testimony of Mr. Wallace Dowd, Transcript, p. 285. "In the TNS system they have what they call a message switch. But it also performs many functions that are traditionally more on the order of data processing specifically. Most of what is in there are identical functions to what we have always had in 3704 and 3705 Communication Controller which has been part of the data processing system" (p. 256). See, finally, FCC, In the Matter of New York Telephone Company Tariff No. 800, Dimension PBX—A Petition for Declaratory Ruling Under Relief, Computer and Business Manufacturers Association, June 16, 1978, p. 5: "The entire dimension offering includes input terminals, printers, display terminals, and a central computer with memory, storage, and coding, sorting and retrieving capabilities."

10. E. Holmes, "FCC Defines 'DP' As Difference Between Smart, Dumb Terminals," *Computerworld*, February 28, 1977, p. 6. See also "AT&T Dataspeed 40 Tariff Stayed," *Electronic News*, February 2, 1976, p. 8; and FCC, In the Matter of AT&T, Petition for a Declaratory Policy That Advanced Communications Services May Be Provided Using Digital Facilities Heretofore Authorized by the Commission," 1978. Bell states: "A restricted view of regulated communications requirement in addition to eliminating the common carriers could well preclude the Bell System Companies from providing needed services to the public because of agreements contained in the 1956 Consent Decree" (p. 2 of the letter from the FCC to James Billingsly).

See also "AT&T's Daring Move into IBM Territory," *Business Week* (July 1974): 48. FCC, In the Matter of Amendment of Section 64.702 of the Commission's Rules and Regulations (Computer II) Docket No. 20828, Reply Comments of AT&T, October 17, 1977: "AT&T recommends a modification of the Commission's proposed rule to affirm that common carriers may engage in the furnishing of processing including data processing in providing communications services within the meaning of the Communication Act." Finally, P. Hirsch, "AT&T's Big Plans for a Value Added Service," *Datamation* (January 1976): 100, defines communication processing as network control, speed conversion, terminal polling, error control, message routing and rerouting, formatting, editing, and so on.

11. FCC Computer II Inquiry, Tentative Decision and Further Notice of Inquiry and Rulemaking, July 2, 1979: "The possible effect of the decree may have on AT&T's ability to offer certain types of equipment and services is a factor to consider in reaching a final decision. Our basic premise is that the Consent Decree should not constrain this Commission in its adoption of regulatory policies which are in the public interest."

12. M. Zientara, "Publishers Decry Bell Home Service," *Computerworld*, December 15, 1980, p. 1.

13. M. Nichols, "Bell, Tex PUC Did 'Irreparable Harm'," *Management Information Systems News,* June 10, 1981, p. 6.

14. E. Holsendolph, "Key to Legal Status of Telephone Accord," *New York Times,* January 14, 1980, p. 1; also A. von Auw, *Heritage and Destiny* (New York: Praeger, 1983), p. 29. Charles Brown, AT&T chairman, described the 1956 Decree as "a fence with a one-way hole in it."

9

THE 1982 CONSENT DECREE

In resisting even partial entry, the Bell System found itself burdened by a rash of private antitrust suits. The government's case was gaining credibility, congressional legislation had reached a stalemate, a 1956 Consent Judgment inhibited Bell's diversification, technology assaulted industry's geographic lines, and an environment of intensive competition was emerging as a new market reality.

On January 8, 1982, a consent decree settled AT&T's longstanding antitrust dispute with the Department of Justice. The agreement submitted to Judge Harold Greene, the presiding district judge, enlisted some 600 responses by interested parties and intervenors.[1] By early August of 1983 Judge Greene modified the original decree, and by the end of the month both parties assented to the judge's final order. With some exceptions, subsequent modifications embraced the essentials of the January agreement between AT&T and the Department of Justice.

The decree alters Vail's concept of the Bell organization and, to a lesser extent, the premise of government regulation. Under the modified decree AT&T will retain ownership of its Long Lines Division, Western Electric, and Bell Telephone Laboratory. Toll telephone service will continue to be regulated by the FCC. But AT&T will spin off twenty-two Bell operating companies, a divestiture of some $80 billion in assets, one-third of its revenues, and nearly 80 percent of its employees. AT&T will relinquish the use of the AT&T name. In return, the 1956 Consent Decree will be lifted, permitting Bell to diversify into fields other than regulated communication common carrier activities.

Under the FCC's ground rules, AT&T may engage in the selling of customer premise equipment and diversify into information management services through the formation of a separate subsidiary. The FCC's nomenclature for such service is termed "enhanced." AT&T has formed a separate subsidiary as a vehicle for domestic diversification, and AT&T International for purposes of exploring international markets.

AT&T will retain its phone stores throughout the United States. However, AT&T will be permitted to enter the electronic publishing services after a seven-year moratorium.[2] Judge Greene imposed the delay in order to permit the growth of this industry—a delay arising from AT&T's prospective ownership of communication facilities and information content.

AT&T's separate affiliate has announced several services: Advanced Information Services Net I (a network permitting the access of incompatible terminals not unlike the old ACS); and a voice storage service known as Custom Calling II. The original Advanced Communication System permitted incompatible terminals to access some 84 percent of all outstanding terminals in the marketplace.[3] Future generations of so-called enhanced services are also in the offing. Presumably, Western Electric and Bell Telephone Laboratory can diversify into information products and services àpropos the market thrust of AT&T. Judge Greene denied a proposal that Long Lines be required to form a separate arm's length affiliate of AT&T.

The twenty-two Bell operating companies will retain control over basic telephone exchange service in their respective territories. About 7,000 local exchanges will be replaced by 161 local access transport areas (LATAs) subject to state commission jurisdiction.

The operating companies will also retain ownership of the yellow pages and after the final asset transfer (1984) offer customer premise equipment for sale. In the future both Bell operating companies and AT&T will emerge as direct competitors in this equipment market.

The regional operating companies, however, are prohibited from manufacturing telephone equipment or in entering the long distance interstate market. The operating company will be organized around seven regional holding companies. Through what is termed a Central Staff Organization, the holding company will engage in research, equipment evaluation, and system specifications and standards for the operating companies.[4] To the extent that this new 8,000-employee operation will engage in research, the centralization of telephone R&D begun by Vail will have moved closer to the operating company level.

Separation of the Bell operating companies from AT&T alters the longstanding toll subsidy employed to contribute to exchange service rates. Presumably, the Bell operating companies will replace that subsidy by imposing an access charge on all long distance companies, AT&T and non-AT&T carriers, that seek to use the subscriber facilities through a disaggregation of interconnection charges. The judge has ruled that the former Bell operating companies must apply access charges equitably to all long distance carriers, Long Lines as well as other common carriers, on a cost-justified basis.[5]

The Bell/Department of Justice Consent Decree constitutes a massive realignment of Bell's assets and markets. That the transfer is both complex and controversial can be seen by the dispute over transfer prices. State regulators have sought to value AT&T's assets at market value whereas Bell seeks to purchase those assets at book value. Judge Greene has ruled that current rate proceedings, namely, book value, constitute the appropriate determinative values of the assets between and among the Bell operating companies.

Rate-making at the state level will never be quite the same again. Not only will the access charge be assigned to individual components, but many expect local exchange rates to trend upward. Flat rate charges are expected to give way to measured rate services in the determination of local telephone exchange rates.

How Western Electric and Bell Telephone Laboratory acquit themselves in this future environment remains uncertain. The transition from monopoly to a competitive culture will not be without trauma. Obsolete and aging plants are being shut down or disbanded.[6] Nevertheless, domestic and international market opportunities are inviting. Western Electric will not only find itself competing with a host of firms for Bell operating company equipment but can also regard the non-Bell equipment market in the United States as fair game. The range of equipment and services overseas may include telephone equipment, computer software, computer hardware, and networks. AT&T is positioning itself for the European Common Market via the route of acquisition and joint ventures.[7]

Western Electric no longer enjoys a complete R&D subsidy as in the past. Under the old license contract, Bell operating companies' subscriber contributions, some $1 billion of revenues to AT&T, carried part of Bell Telephone Laboratory's research expenses. Except for the contributions of the Long Lines Division, Western Electric will bear a heavier burden of R&D funding, an expenditure that will be reflected in Western Electric's prices.

Only time will reveal how the regional operating companies, AT&T and Western Electric, and Bell Telephone Laboratory will fare in this new environment. Not unlike the 1956 Consent Judgment, the 1982 decree rests upon the existence of demarcations of industry, product, and service boundary lines. The domestic market, for example, is obviously premised on a separation of local exchange service and toll or long distance service; the local service decreed a natural monopoly and the toll was open for competition. But how long will exchange revenues remain impervious to market entry? After all, technological substitutes abound in equipment, in inside wiring, in local area networks, and even in local loop distribution systems, to say nothing of digital switching. The number of firms moving as potential candidates and offering a variety of services continues to expand and multiply.

AT&T may circumvent the Bell operating companies' local distribution facilities along with, of course, the competitors of Long Lines, including satellite companies. What constitutes the local loop may no longer be a matter of manifest clarity, nor will demarcations between toll and exchange service remain static, to say nothing of the classic separation of communications, information, or broadcasting.

Nor will domestic and international market distinctions be as sharp as in the past. While it is true AT&T is expanding into overseas markets, offshore suppliers are entering the United States, positioning themselves for new markets and new product opportunities. (Witness the United Kingdom's Plessey Corporation's acquisition of Stromberg-Carlson from United Technology.)[8] That Western Electric is certain to view its future direction in global rather than domestic terms can be seen from the company's observation: " . . . we are taking some strong steps ourselves to internationalize our product lines so that it fits the world."[9]

The 1982 Bell Consent Decree represents a massive distribution of assets, an unprecedented rearrangement of markets and services, a realignment of domestic and overseas markets. But how long will the presumption of industry boundary lines remain valid? Will the 1982 Consent Decree be as flawed as the 1956 judgment through the imposition of static, orderly demarcations? Will the premise of market distinction be eroded by technology? Has the state of the technical art rendered the decree obsolete as the ink dries? These questions remain unanswered. In the meantime, the institution of public utility regulation addresses a new mandate. What, in short, is the direction of regulatory oversight?

NOTES

1. United States v. American Telephone and Telegraph Company, Western Electric Company, Bell Telephone Laboratories, Civil Action No. 74-1698, U.S. District Court, District of Columbia, August 11, 1982, p. 2.

2. Ibid., p. 99: "The electronic publishing industry is still in its infancy." See also K. Jones, "Texas Publishing Execs Hail 7-Year News Ban on ATT," *Management Information Systems Week,* September 1, 1982; and M. J. Richter, "Judge Reverses Ban on Telex Equipment Marketing," *Electronic News,* August 16, 1982, p. 1.

3. A. Pollack, "AT&T to Enter Data Field," *New York Times,* June 16, 1982, p. 34.

4. M. J. Richter, "Blast Procurement Center, Licensing Restriction in Bell Reorganization Plant," *Electronic News,* February 21, 1983, p. 1.

5. I. Stelzer, "The Post-Decree Telecommunication Industry," *National Economic Research Associates, Inc.,* Princeton, N.J., March 11, 1982.

6. Warren Brown, "Western Electric's Modernization Means Pain for Baltimore," *Washington Post,* January 31, 1983, p. 17. "The terms abound: obsolescence, overcapacity, redirection, competition. But for Western Electric Co. nowadays, they all mean one thing—pain"; Bob Vinton, "WE Plants' Fate May Hinge on Plans for Merchant Market," *Electronic News,* February 14, 1983, p. 50; Robert Hanley, "Kearny Plant is Dying Along With an Old Era," *New York Times,* January 29, 1983, p. 25; B. Levine, "WE to Ax 6,700 In Plant Cutbacks, Take Loss for QTR," *Electronic News,* January 31, 1983, p. 1; Robert Vinton, "WE Set to Close No. Illinois Works, Affecting 1,250," *Electronic News,* August 16, 1982, p. 20; Mark Potts, "Fading Shadow, Western Electric's Hawthorne Works Fights to Survive," *Washington Post,* January 2, 1983, p. F1: "Despite the demolition of many of the plant's older buildings, despite the construction in recent years of more modern facilities on the site, Hawthorne is still a fairly old manufacturing plant, little suited to many high technology chores." "My buildings are old and drafty, the sprinkler systems let go every once in a while because the pipes are old and you have to patch the pipes."

7. P. Lewis, "Bell Flexes a Muscle in Europe," *New York Times,* September 24, 1982, p. 33.

8. "Plessey Makes a Good Connection," *Financial Times,* September 27, 1982, p. 7.

9. P. Betts, "U.S. Telephone Service: Plain and Simple It Won't Be," *Financial Times,* January 12, 1983, p. 12. "We are taking steps to internationalize our product line so that it fits across the world" (statement by C. Brown, AT&T chairman). See also "France Is Disconnected in an AT&T-Philips Link," *Business Week,* October 11, 1982, p. 47; and "AT&T Says It May Look Abroad to Offer Some Debt and Build Production Plants," *Wall Street Journal,* January 20, 1982, p. 8.

10

FEDERAL REGULATION

In its fifty-year history, the FCC has grown in power, visibility, and influence. As an arm of Congress, the commission stands as a quasi-judicial regulatory body. Nevertheless, it prizes its independence from the executive, judicial, and legislative branches of government. The commission is a critical agency in determining the direction of U.S. information policy.

In a broad sense, the FCC has led two lives. In the first half of its institutional life, it concentrated on defending natural monopoly from technological or competitive intrusions. During this era, the commission adopted the structure, practices, and policies of the telephone industry. For the last twenty years, however, the commission has adopted a pro-competitive stance. In this second era, a period of discretionary market entry, the commission began to challenge the telephone industry's policy and structure and to experience conflict and controversy.

MARKET ENTRY

A policy that aids and abets competitive entry in a world subject to regulated markets at first glance appears to be contradictory. Either a firm enjoys the efficiency of a monopoly status or it does not. Can a market afford a regulatory halfway house—a segmentation where diversity and exclusivity work to the benefit of the same consumer and yet a segmentation where market exclusivity works to the benefit of the user? During the past twenty years, the FCC affirmed and embraced this segmented view of its mandate, much to the distress of the telephone industry and to the opposition of state commissions. Such a policy change found the Bell System and the FCC arrayed as protagonists, a situation that prompted AT&T to seek congressional relief to transfer critical FCC authority to state jurisdictional control, an effort that failed in the mid-1970s.

Although no one decision marks the point of departure, the FCC's decision to relax frequencies for private microwave point–to–point communications in 1959 and 1960 represented an early break in traditional policy.[1] The telephone carriers strenuously opposed this limited form of access. Users and suppliers of microwave radio equipment obviously supported a relaxation of frequency use. When the commission voted to permit user ownership of radio relay, the FCC launched a policy that was antagonistic to the carrier's view of its regulated market. In a word, the commission broadened the options confronting large corporate and government users of communications.

That the Bell System was upset by this departure from past commission policy is an understatement. Under commission policy, no telephone carrier need interconnect its telephone switched facilities to the customer's private microwave link. Any firm opting to build such a private system was constructing an expensive intercommunication system restricted to intracompany application.

Railroads were particularly upset that in the event of a bridge washout the call from the engineer's cab could not access the home of the individual supervisor.[2] Under FCC interconnection policy, the engineer would be required to locate a telephone pay booth. Although AT&T relented by permitting emergency access in the railroad case, interconnection of private and carrier facilities was banned as standard carrier practice FCC policy. Users who elected to construct microwave links were thus denied access to interconnection to the public switched network. Not surprisingly, the microwave potential for equipment and services never developed.

The late 1960s and early 1970s witnessed a further step in the commission's segmented market approach. In 1963, an Illinois company, MCI, applied for a franchise as a specialized carrier between Chicago and Saint Louis. Six years later, by a margin of one vote, the FCC approved MCI's application for service.[3] Inundated with dozens of specialized carrier applications, two years later the commission sanctioned entry into the private lease market on a come-what-may basis.[4]

The precise dimension of what made up leased communications services was somewhat unspecified at the time. Was private line service to include only point-to-point service, forward exchange service, and corporate switching services such as Bell's? Whatever the answer, the commission sanctioned entry with the requirement that local telephone companies make their distribution facilities available to new specialized carriers.

A year later, the commission, under White House prodding, adopted an open entry into the domestic communications satellite market. In promoting such a policy, the commission had come a long way from its original decision to restrict satellites to international carriers and international communications.[5]

MARKET DIVISIONS

Once entry was approved, the FCC sought to establish and adjudicate clear distinctions between monopoly markets and competitive markets. To the extent that markets overlap, the FCC attempted to orchestrate and balance a blend of market order and market diversity. From 1960, the FCC postulated market definition and boundaries. At first glance, the task appeared eminently achievable. Ambiguity was to be eliminated, and clarity was pursued as a virtue. But as technology accelerated, what began as a relatively simple endeavor became increasingly complex. The difficulty was that once the ground rules for entry were established, technology eroded the market lines, forcing the commission to start another investigation and to impose new rules for market segmentation.

In some respects, the commission's attempt to segregate markets was inherited from the telephone and broadcast industry. The 1934 communication act distinguished between the two industry segments on the premise that the telephone and telegraphy would not mix. With this division as precedent, the commission's future policy direction might well have been predicted.

In the 1950s, for example, the FCC perceived any market overlap between AT&T and Western Union not as healthy rivalry but as economic disorder. This overlap was a legacy of an era when both companies were direct rivals; Western Union owned telephone companies, and AT&T owned telegraph companies. In 1951, the FCC approved a mutual swap.[6] AT&T was now confined to message toll telephone and Western Union to public message telegraph service.

Communication satellites placed a special burden upon the FCC's ability to delineate markets. In the early 1960s, U.S. aerospace firms, General Electric, Lockheed, and others sought permission to offer satellite service overseas on a commercial basis. The FCC first divided the market geographically into domestic and international services. From this distinction, the commission allocated overseas satellite use to the international telephone and telegraph carriers, thus blocking market entry to GTE, General Electric, and Lockheed.[7]

This market segregation caught some international carriers off guard. The staff of the Small Business Committee pleasantly surprised an overseas subsidiary that it was not merely an international regulated carrier subject to FCC jurisdiction. It was now part of a consortium to establish international commercial satellite service.[8] To their dismay, General Electric and Lockheed found that they had been frozen out of participating in commercial satellite service overseas. The satellite dispute eventually erupted before Congress, and in 1962, through the Communication Act, COMSAT emerged as a joint venture.

The FCC continued its tradition of postulating market distinctions through the mid-1960s. Both AT&T and Western Union offered a competitive teletypewriter service; AT&T's was called TWX and Western Union's Telex. To reinforce Western Union's economic viability, the Common Carrier Bureau and eventually the FCC approved the telegraph union's acquisition of Bell's TWX under a joint TWX/Telex service.[9] The goal of this policy was to impose a line between message telephone service and public message service or public telegraph service.

As technology began to nibble away at traditional demarcations, however, the commission's endeavor to identify and segregate borders became increasingly frustrated. If a firm programmed a computer to both process and route information by a computer terminal over telephone lines, what precisely was the content of that service and what was the status of the firm offering that service? Was the service communication or data processing, or both? And was the firm a regulated common carrier, or an unregulated competitive firm, or both?

This case proved more than hypothetical. In the 1960s, Bunker Ramo sought to computerize a stock quotation service employing terminals, communication lines, and a central computer.[10] Both Western Union and AT&T refused to make available leased communication facilities to Bunker Ramo on the grounds that the firm was engaged in a communication-regulated activity. Yet, the firm possessed no FCC license.

When Bunker Ramo turned to the FCC for resolution, the commission served as a broker between the carriers and the data processing firm. Eventually, Bunker Ramo agreed to eliminate administrative messages from its service, and AT&T consented to provide communication facilities in order to enable the company to establish a computerized stock retrieval service. Still, it was obvious that any computer firm could easily program a mainframe to process and route information simultaneously. Technology had penetrated the essence of a common-carrier service, message switching.

The struggle over definition, however, resisted a one-time solution. In 1961, the FCC issued several proposals with regard to combining the computer and communications.[11] If a firm's service was defined as communications, the FCC held that service was subject to regulation. If the service was defined as data processing, the commission would forbear regulatory control, even though it clearly possessed regulatory jurisdiction. But if a firm offered data processing and communications simultaneously, a so-called mixed or hybrid service, the mix of the service would determine its regulated status.

It was this amorphous, shifting area that proved most disconcerting and frustrating to regulation. If, for example, the communication component exceeded the data processing component, regulation would be invoked. On the other hand, if data processing exceeded communication, regulation would not be extended to the firm. How could the FCC or, indeed, anyone balance the shifting ratios of computer/communication services? How could regulation monitor dynamic boundary changes with precision? In 1971, the commission, invoking what it termed the primary business test, attempted to do so by rank-ordering the two computer/communications services.[12]

In the meantime, the migration of microelectronics from computers to peripherals and from terminals, displays, and minicomputers to storing devices continued to be driven by the declining cost of logic and memory. These devices became endowed with storage, buffering, and processing capability. The distinctions between computer mainframe, computer terminal, and computer peripheral began to soften and erode. And the FCC's market distinctions, promulgated in 1971, were soon bypassed by events.

No sooner had these definitions been posted than the FCC instituted another set of rules for computers and communications. At first, the commission proposed to distinguish between arithmetic processing, word processing, and processing controls, lumping them under a policy called "regulatory forbearance." On the other hand, the FCC proposed to regulate what it called "network control and network routing" as elements essential to communication switched common-carrier services.[13]

Later, the commission offered three additional classifications (what it termed "enhanced communication, enhanced data and pure data") as one way of distinguishing submarkets subject to regulation and to competition. Still later, the FCC collapsed these categories into two broad groups: basic communication services and enhanced services. Basic services were subject to regulation and communication services to competition. But nowhere was the regulatory frustration more apparent than in the agency's attempt to define terminals. A new teletypewriter Model 40, introduced by AT&T and manufactured by its manufacturing affiliate, underscored how quickly definitional boundary lines would change.

The FCC classified AT&T's Model 40 terminal as a communications device appropriate for tariff filing and regulation. The exercise was not without strenuous effort. The Common Carrier Bureau of the FCC attempted to segregate terminals on the basis of microchip population. If a device excluded chips, it was labeled "dumb." If the terminal possessed a microprocessor, the device was labeled "smart."[14] With this distinction as its operating premise, the Common Carrier Bureau pronounced the Model 40 terminal to be a "smart" terminal and stated that it was beyond AT&T's lattitude to offer this service to the public. The bureau reasoned that a "smart" terminal constituted data processing, and, by the AT&T Consent Decree of 1956, the firm was not able to offer this service to the public.[15]

Market segmentation also erupted amid joint ventures and acquisitions. When IBM acquired a satellite firm, CML (a joint venture of Lockheed, MCI, and COMSAT), the FCC approved the purchase on certain conditions. The FCC determined to separate computers from data processing, specifying the number of firms to join in the joint venture and the degree and type of transactions occurring between IBM and Satellite Business Systems, the new satellite venture.[16]

When General Telephone and Electronics acquired a value-added carrier, Telenet, the FCC once again sought to establish an ambiguous demarcation between GTE, the parent, and Telenet, the operating affiliate. Telenet, as a value-added affiliate, was subject to guidelines, including financing, R&D, transactions, and marketing.

All this served as a prelude to AT&T's activities. When Bell sought to introduce satellites into domestic communications, the FCC limited satellites to Bell's monopoly services, imposing a moratorium of several years on AT&T's use of satellites for private leased services. The FCC's effort to classify, assign, and segment markets can be seen by the proliferation of definitions issued over the past fifteen years.

- Telephone versus telegraph
- Private line versus message toll telephone
- Data processing versus communications
- Resale versus sharing
- Smart terminal versus dumb terminal
- Wire line carrier versus common carrier
- Broadcast versus common carrier
- Enhanced versus basic service
- Message service versus circuit switching
- Hybrid data versus hybrid communications
- Arithmetic processing versus electronic processing
- Terrestrial microwave versus satellite microwave
- Basic voice versus basic nonvoice
- Dominant carrier versus nondominant carrier
- First tier company versus second tier company
- Bell equipment versus general trade supplier equipment
- Cable TV versus broadcasting firm
- Point-to-point private line service versus switched private line service
- Basic voice versus basic nonvoice versus enhanced service
- Domestic satellite versus international satellite
- Process control versus data processing, and so on.

The FCC issued an impressive catalogue of semantics, drawing finer and closer distinctions, sanctioning entry into some markets, banning entry into other markets, and promising to look at still others. In the process, the commission found it necessary to extend and broaden its jurisdictional overview. Market entry, therefore, served as a policy end; jurisdictional drift evolved as an institutional means.

JURISDICTIONAL DRIFT

Jurisdictional drift has been as subtle as it has been inexorable. What began as authority over telephony progressed to supremacy in the interconnection of specialized carriers at the state level and to lifting of the Department of Justice Consent Decree over AT&T. Jurisdictional creep did not occur in one giant step, but rather through a series of incremental steps, all of which were promulgated under the public interest doctrine.

The evolution began with the telephone set. The telephone set performs a dual function: local exchange calls and long distance calls. The set thus straddles two jurisdictions—state and federal. In 1968, when the commission ruled that customers could buy and attach their equipment to the dial-up network, state utility commissions objected that their regulatory prerogative had been transgressed. Upon judicial appeal, the courts upheld the primacy of FCC jurisdiction.[17]

The courts upheld the commission's determination to tariff Bell's computer terminal. The appeal court asserted that the FCC operated within its jurisdiction and that its function was not to second guess the expertise of a regulatory body: "Petitioners (IBM) challenge the FCC's classification in an attempt to second guess a specialized agency's application of its own rules to stipulated fact. Mindful of the reference to agency conclusions and such circumstances we are not disposed to intervene."[18]

Specialized communication interconnection represented a similar jurisdictional drift. As firms sought to enter the private lease market, access to local telephone facilities emerged as critical. To ensure such access, the FCC asserted jurisdiction over local connection, despite the opposition of state public utility commissions. This jurisdictional extension was again upheld under judicial review.[19]

In the mid-1970s, when firms sought to lease and resell telephone lines to different customers, the FCC broadened its jurisdiction to include resellers, despite the fact that the capital invested in transmission facilities resided not with the resellers but with the original common carriers. The commission reasoned that reselling fell within its jurisdiction because the firm contemplated a markup to secure the return on their investment in enhanced services.[20]

In data processing and communication services, the FCC's jurisdictional drift picked up speed in the 1970s. In 1971, the commission stated that it held regulatory authority over companies using computers over telephone lines but declared it would not exercise its jurisdictional authority. Regulatory forbearance evolved as an extension of authority but was held in abeyance, according to the commission. On the other hand, when the commission recommended that non-Bell companies or carriers form a separate subsidiary for data processing diversification, the commission imposed a transaction ban between the operating affiliate and its regulated parent. The courts ruled in this case, however, that the FCC had overstepped its antitrust jurisdiction.[21]

As the commission witnessed AT&T move into data terminals and protocol conversion and as the commission struggled with semantic complexities, it became clear that Bell's 1956 agreement posed a formidable barrier to technological diversification. Soon, a problem that involved Bell management became a problem adopted and "owned" by a regulatory agency. AT&T's 1956 Consent Decree now had emerged as a regulatory issue.

At first, the commission accepted the decree's premise as a constraint to Bell's diversification into data processing, a stance which the commission adopted in 1971. Five years later the commission reconfigured the definition of communications so as to include the Data-speed Model 40. Finally, the commission lifted the decree entered between the Department of Justice and AT&T in 1980. In its Computer II finding, the FCC concluded, "Our basic premise is that the Consent Decree should not constrain this commission in its adoption of regulatory policies which are in the public interest. . . ."[22]

RASMUSON LIBRARY
UNIVERSITY OF ALASKA-FAIRBANKS

The commission ruled that Bell could diversify into services "incidental" to common-carrier activities and that AT&T must establish a separate affiliate for nonregulated information services. The FCC will, of course, monitor what constitutes communications and information content appropriate for Bell's new affiliate. In the process, the following services are subject to regulatory observance:

- Distributed data processing
- Word processing
- Electronic mail
- View data
- On-line bibliographic services
- Home information systems
- Electronic funds transfer
- Home computer systems

When queried as to any possible leakage between Bell's regulated and unregulated affiliate, the commission observed that its "ancillary jurisdiction" over the nonregulated firm would best serve the public interest. Hence, the FCC will monitor information, however defined, as Bell diversifies and establishes new services.

That definitional opportunity occurred sooner than later. The U.S. Post Office embarked on an electronic mail program called ECOM, a message sent by terminal and satellite to a post office, and then hand delivered to the ultimate residential or office destination. The FCC asserted that communication firms providing facilities to the post office had to file tariffs before the commission and, that to the extent that the post office employed communication facilities, it would be classified as a common carrier subject to the commission's jurisdiction. The issue is now in the courts.[23]

The FCC has been ambivalent as to its role over vertical integration. As part of the 1934 Communication Act, Congress ordered the commission to investigate whether competitive bidding ought to be imposed on Bell's purchases of equipment, services, and facilities. A 1938 report by the commission did request enabling legislation so as to give the commission direct regulatory authority over Bell's manufacturing affiliate.[24] The 1938 report insisted that Western's prices affected AT&T's rate base, a cost ultimately borne by rates paid by the subscribing public. The report also insisted that an enlarged jurisdiction of the equipment was essential to the determination of fair and reasonable rates.

Congress resisted this jurisdictional authority, and a year later, the FCC instituted a policy known as continual surveillance while asserting that legislation was not needed to give the commission jurisdiction over all aspects of the telephone industry.[25] By the middle 1970s an FCC investigation by a special trial staff recommended that the commission limit its jurisdiction over product manufacture. It recommended that the FCC report to Congress that vertical disintegration was in the public interest and that the tie-in between manufacturing and affiliate acted to foreclose the equipment market.[26] The trial staff recommended that the ball be placed in the congressional court.

Finally, the FCC has not hesitated to assert its jurisdiction over structure proposals engaged by the Department of Justice in its antitrust suit against the Bell System. The FCC reminded the Department of Justice of its pervasive and regulatory authority over the telecommunications industry. When the Department of Justice proposed to spin off twenty-two operating companies from AT&T, the FCC asserted that its control of radio licenses gave it jurisdiction over the selling of assets and a general overview of Bell's structuring.[27] The commission suggested a Section 214 investigation, reassuring the Department of Justice that all parties would come forth and make their comments to be assessed by the FCC's mandate to oversee the public interest.

POLICING BOUNDARIES

Once entry into specific markets is encouraged, and once jurisdiction is asserted, regulation must monitor the boundary lines to ensure the integrity of demarcation. Stated differently, regulation must police demarcations. Over the past decade there have been four attempts at borderline monitoring: MCI's Execunet service, the pricing practices of AT&T, the manufacturing of telephone equipment in the United States, and the dispensation of exchange subsidies.

Execunet

MCI's original service opposed point-to-point transmissions by its investment in microwave relay equipment. But the evolution in computerized switching became attractive by the mid-1970s. Customers of private line circuits could hook up to computer programs to route, store, and transmit voice, data, and documents to several geographic locations. A user needed only to buy a computer off the shelf from suppliers either at home or overseas.

Not surprisingly, MCI purchased such a device from a Canadian firm and offered a computerized switched service to toll subscribers. AT&T insisted that MCI had illicitly entered into a wide area telephone or message toll telephone market without explicit FCC permission. The viability of any toll service, of course, rested upon MCI's ability to gain access to local telephone exchanges, and here Bell's control of the local operating companies proved critical. AT&T restricted local access to MCI's Execunet service, thereby arresting the service's growth.

Similarly, the FCC viewed MCI's Execunet service as being in violation of the ground rules issued in permitting MCI to go into business. Accordingly, the FCC ordered AT&T to deny MCI access to Bell operating companies and their local exchange facilities. This action temporarily killed the company's Execunet service.[28]

MCI took AT&T and the FCC to court, and in a surprising decision the appeals court overturned the commission's sanction against MCI.[29] Subsequently, the commission promulgated a local access charge. Specialized carriers today offer a switched long distance service to their customers.

The FCC's attempt to penalize a new carrier was ironic. The commission found itself restraining MCI from diversifying from point-to-point communications services while at the same time lifting an antitrust consent decree to permit AT&T to diversify beyond regulated services. The supreme irony was that AT&T, in placing itself in the regulatory box, had viewed the 1956 Consent Decree as an exercise in management perspicacity. Now Bell supported FCC efforts to break out of its own box.

Rates

Once regulated markets are established, they must be monitored and policed. But a firm that simultaneously occupies a monopoly and a competitive market presents an elusive target. Such was the case when Bell rendered a monopoly MTS service on one side and yet competed with the specialized carrier in either a private or switched private line on the other side. How could regulation protect services insulated from competitive access from underwriting losses in competitive markets, thus denying the reality of competition over the long term? How could the FCC prevent rate subsidization that would render null and void its attempt to introduce segmented competition? The answer focuses upon questions of rate structure and cost allocation.

The FCC and AT&T have explored alternative pricing/costing methodologies for nearly twenty years. Bell insists that, while a fully allocated cost standard is appropriate for some monopoly services, an incremental or marginal cost standard for competitive services is also legitimate. The FCC's position is that two separate cost standards is tantamount to cross-subsidization and will lead to a diminution of competition in one segment of Bell's overall market.

After fifteen years, the FCC adopted a compromise rate structure known as fully distributed costs -7 or FDC -7. This cost policy embraced elements of past and future cost estimates. Next, the commission embarked on a program to implement, execute, and apply those standards and to collect data in order to establish prices for particular communications submarkets.

Given a plant of over $1 billion that straddles two markets, even under the best of circumstances, cost allocation is an arbitrary undertaking and a formidable exercise. One FCC employee describes such adventures as a "trip to the center of the earth."[30] Along the way, the FCC ruled that Bell's competitive tariffs were discriminatory or lacked cost documentation. The FCC ruled that each tariff (Telpak, Series 1100, Hi-Lo, MPL, DDS, and DSDS) suffered from inadequate data and cost information so as to render impossible judgments of reasonableness. In frustration, the court imposed on the FCC's search for a cost allocation formula. Later that commission abandoned its fully distributed costs/seven and adopted an interim cost manual as a short-term pricing expedient.[31]

The commission's attempt to monitor cross-subsidization between two markets is frustrating and costly. Telpak, a tariff imposed in response to private micro-wave competition, was litigated for almost twenty years before its retirement.[32]

Promulgating economic standards is lengthy, time-consuming, and elusive. The question remains, can regulatory policies determine submarket prices and costs with precision and accuracy? The General Accounting Office (GAO) appears to think that the issue will be resolved if the FCC hires seventeen additional staff members.[33]

Protecting an Equipment Cartel

The 1956 Consent Decree did not ban Western Electric from selling equipment to the independent or non-Bell telephone companies. Rather, the decree inhibited Western from engaging in noncommunication-related activities. As a matter of practice, Western Electric withdrew from selling to the independent or non-Bell operating companies after 1956 (except in exceptional cases). A tacit cartel thus evolved between Western Electric and non-Western suppliers. Some dozen companies competed for access to the USITA or non-Bell telephone market, while Western enjoyed privileged access to the twenty-two Bell operating companies.

The FCC has viewed the equipment market as peripheral to its central agenda of broadcast and common-carrier activities. Vertical integration generally perplexes the commission. The FCC did attempt to impose direct regulation on Western Electric in the 1930s; but this effort was abandoned as too radical even during the depression. To the extent that equipment costs translate into telephone rates, the commission possessed no market benchmark to assess Western's performance, productivity, and efficiency.

Perhaps this explained why the commission never disallowed Western's prices from AT&T's investment rate base. Whatever the answer, indirect regulation of equipment production posed as one policy option. But by electing neither competition nor monopoly in the equipment market, the FCC opted for essentially nothing. As a result, the FCC has presided over an inherited cartelized equipment market for the last fifty years.

By the mid-1960s, an activated FCC decided once again to look at Western Electric and AT&T's prices and costs. The endeavor was long and costly. After eleven years, 1 million pages of documents, over 100 trial days, 16,000 exhibits, an expenditure of over $5 million, and an equal amount spent by AT&T, the commission concluded that neither AT&T's ownership of Western Electric nor the operating companies' purchase of 90 percent of their equipment from Western Electric constituted "a convincing showing that Western used its position in the Bell System to get an inside track on sales."[34]

An administrative law judge held that Western Electric and AT&T epitomized the essence of market performance and recommended that Western be unleashed to compete with the competitive non-Bell or independent telephone company industry.[35] The administrative law judge insisted that AT&T should permit Western Electric to compete for sales with Rochester Telephone, Lincoln Telephone,

General Telephone, United Telephone, Mid-Continent, and all independent telephone companies.

Suppliers of the independent telephone industry were perplexed by the judge's finding. Foreclosed from competing with Western Electric's access to Bell operating companies, the judge recommended that Western Electric compete for access to non-Western customers. Nevertheless, the full commission admonished AT&T to broaden access to its equipment market via a centralized buying entity. Before Docket No. 19129 was ended, AT&T established a Bell Systems Purchase Product Division (BSPPD) to facilitate access by independent suppliers to Bell operating companies.[36]

But the BSPPD was obviously beset by inherent conflicts. Its parent, AT&T, owned Western Electric, and yet BSPPD was expected to promote the fortunes of Western Electric's competitors. A BSPPD employee alleged that the operating companies all but ignored AT&T recommendations and continued to buy equipment exclusively from Western Electric.[37]

Some suppliers, acknowledging the longstanding relationship between AT&T and the Bell operating companies, suggested that outside general trade suppliers be assigned buying quotas as a short-term expedient. These manufacturers argued that a quota system would force Bell operating companies to purchase non-Western Electric hardware.[38] The FCC demurred, however, and rejected this quota plan while reaffirming AT&T's ability to impose quotas on its own purchases.

Bell's decision to divest twenty-two operating companies marks a partial disintegration of utility and supplier. The Bell operating companies are free to sell equipment, including terminals, and to purchase a variety of telecommunication hardware and products. That vertical integration persists in the AT&T/Western Electric relationship is seen by the commission's fiber optic cable decision.

An application of exotic techniques to pure silica fiber optics enjoys a thousand-time capacity increase over coaxial cable. Fiber optics can stretch as far as 100 miles without repeaters, in contrast to a 1-mile repeater for conventional coaxial cable systems. Fiber optic cable lends itself to intercity trunking, and some overseas companies are employing optics for local distribution facilities.

Bell's Northeastern U.S. fiber optic cable link served as an important commercial application. Despite Bell's BSPPD and despite the presumption that general trade suppliers could get access to AT&T's market, a Bell official nevertheless observed that on fiber optics: "We have never seen the need for bidding in the past."[39] Western Electric's optical cable award prompted Corning Glass to file a complaint with the FCC. The commission then asked Bell to solicit competitive bidding on fiber optics. Some ninety pages of specifications on fiber optic equipment were issued, and Bell solicited tender offers from eight firms, including Western Electric. Fujitsu submitted the low bid.[40]

A backlash erupted. One FCC commissioner suggested that Fujitsu's bid was illegally below cost, and Congress expressed concern that a Japanese firm would secure a domestic cable award. AT&T withdrew its award and on grounds of "national defense" assigned the contract to Western Electric.[41] The company did not elaborate as to the nature of national security. Whatever the rationale, the FCC observed that since Western Electric subcontracted components to other U.S. suppliers, such a policy would "provide a more competitive telecommunication equipment manufacturing industry."[42] AT&T later announced it would entertain bidding firms if suppliers manufactured cable within the continental United States. The commission since has remained silent.

DISPENSING A NATIONWIDE SUBSIDY

Today the FCC is moving to sustain, feed, and preserve rural telephone companies in the United States through a massive subsidy program. For decades, long distance calls have borne a part of the cost burden of local exchange service that accounts for some 80 percent of telephone exchange costs. Known as toll settlements, AT&T allocated some 30 percent of toll revenues to the exchange market.

The separation of AT&T and its Bell operating companies effectively terminates the toll subsidy. As a temporary replacement to the settlement process, the FCC has introduced the concept of local access fee. The fee is to be imposed on the local telephone user (residential and business) and interexchange carriers, including AT&T and its intercity rivals. The subscriber charge will be billed on a flat monthly basis; the interexchange levy posted on a usage basis. A portion of the carrier charge will be assigned to a revenues pool dispensed to some 1,400 rural telephone companies throughout the United States.[43] In the commission's words, "There is however no reason to believe that regulatory actions sending proper economic signals cannot preserve the long term viability of the nation's local phone companies from suffering debilitation from bypass."[44]

The FCC finds itself once again manipulating prices as a means to inhibit competition in the local exchange area. The access fee imposed on local telephone users presumes FCC jurisdiction now applies to local exchange service as well as interstate telephone service.

This attempt to monitor relative prices turns on the number of available substitutes. Residential subscribers, for example, may possess few short-term options to local services and may be locked into higher rates. Large corporations are not similarly constrained, however. Technology continues to yield new and intriguing combinations of satellites, fiber optics, cable broadcast, and FM radio. It is not inconceivable that competition will erode that tax base premise of the local subsidy over time, particularly as the access fee is scheduled to rise over the next several years.

Business users exploiting loop alternatives for whatever reason—cost, price, band width capability, transmission quality—will prompt AT&T and other long distance carriers to engage in technical bypass themselves. If this development gains strength, the subsidy pool can dissipate through a combination of technological substitutes and market competition. After all, some 60 percent of toll revenues are derived from about 4 percent of business users,[45] and users themselves are evolving into new entrants in telecommunications.

The FCC's Universal Service Pool layers one telephone subsidy on top of another. However meritorious in intent, public policy has yet to assess the economic implications of such subsidies. Do prices set below costs stimulate telephone company efficiency? Do revenue contributions harden a cost plus, capital intensive corporate mentality? Do subsidies perpetuate existing loops, terminals, switching, and investment? Does pricing below cost discourage rural phone companies from exploring digital concentration techniques, alternative tariff, line sharing, and so on; and is it true that " . . . it's now possible—through electronic enhancements—to handle nearly every foreseeable transport capability with existing loop facilities?[46]

The Universal Service Fund institutionalizes a tax/subsidy calculus in an era of furious technological change. Why is it that in attempting to promote market competition the FCC expands its jurisdiction and solicits an additional set of clients dependent upon its revenue largesse? Perhaps the answer is that markets are controlled because " . . . it is an illusion that most intelligent civil servants believe they know best."[47]

SUMMARY

In summarizing, the FCC must be applauded for improving market entry, even though competitive policy has been incremental, hesitant, and partial. Entry of any type of specialized service represents a break with past tradition. But a policy of partial entry is not without its consequences.

First, discretionary entry becomes a poisoned process. An outside firm seeking access reminds a regulatory body of its commitment to expedite innovation, promote efficiency, broaden consumer choice, and advance the state of the telecommunication art. Once a franchise is received, the new firm converts to an incumbent firm. The incentives to align itself against a potential entrant prove irresistible; the firm resists competition on the bases of duplication of facilities, excess plant capacity, revenue cream-skimming, or the economic power of the new entrant. Inevitably, the incumbent firm seeks to protect its franchise from pressures of market accountability.

Regulated entry creates institutional resistance to subsequent market access. Entry no longer becomes a process but an event. Indeed, firms employ due process to retard rivalry. Yesterday's hero becomes tomorrow's obstructionist.

Examples of this poisoned entry process abound. MCI fought the good fight against the telephone industry and then spoke out against truckers sharing private microwave, labeling them semicommon carriers.[48] Telenet packet switching proceeded as an innovative new firm, but the firm protested that Tymnet's data processing was communications and hence subject to regulation.[49] Graphnet creatively introduced a facsimile service but opposed ITT's similar service on the grounds that entry was counter to the public interest. Cable TV firms opposed telephone company pole line restrictions yet sought extension of regulation to satellite master antenna systems (SMA TV). Radio common carriers, struggling to remain a viable force in telephone company areas, now oppose common carrier deregulation of FM subcarrier channels; broadcasters favored microwave video transmission but oppose direct broadcast satellites.

Short-term access sets up the very force that impedes long-term entry. Discretionary entry evolves into a regulated cartel aligned against further competitors. As the process grinds to a halt, cartel supervision becomes more manageable and less chaotic.

Second, the FCC's jurisdictional reach expands relentlessly. Over the years, commission power has superseded the Department of Justice, the U.S. Post Office, the Executive Office of the President, and state regulatory agencies. Recently the commission reclassified remote data processing firms as "carriers" so as to justify imposing higher communication access charges. Does this mean that Sears' Allstate, Hilton Hotels, and the New York Teleport fall within the reach of FCC jurisdiction?

Third, once jurisdiction is reached, the commission divides markets, balances and polices divisions with vigilance. Market boundaries mandate commission oversight; and if firms dare jump their assigned turf, the FCC reserves the right to impose sanctions and penalties.

Fourth, market boundary lines invite a search for rate cross-subsidization. The FCC has sought to post cost allocations and market standards in order to police and prevent rate cross-subsidization of monopoly/competitive markets. Having abandoned traditional accounting techniques, the FCC has created separate subsidiaries. Perhaps this approach will work, but twenty years of rate structure cases have yielded precious little.

Fifth, the commission has served as patron of an equipment cartel without imposing standards of economic performance. When competitive bidding is introduced, the process is withdrawn under appeals to "national defense." Vertical integration remains imbedded with regulatory indecision or indifference.

Sixth, in the name of the public interest, the FCC is evolving into a collector of taxes and a dispenser of subsidies—all of course for a good cause—the rural telephone subscriber. But the opportunity cost of rate intervention has never been calculated nor explored.

Seventh, industry market boundaries continue to be imposed—the latest endeavor "basic versus enhanced." But will such demarcations apply to the separated Bell operating companies, the independent telephone company, rural and metropolitan telephone companies? Will not such boundaries be subject to technological obsolescence as previous endeavors divide and segregate?

Eighth, regulation treats time—a most precious commodity—as essentially a free good. True, the commission expeditiously extends its jurisdictional reach to include depreciation, information, market separation, access charge, and so on. But once jurisdictional turf is secured, the "tar baby" of due process becomes ponderous and time consuming. Inevitably, due process delay invokes an opportunity cost in terms of services, equipment, and choice denied.

In sum, as due process is less able to deliver, regulation becomes "federalized"; as FCC power and authority multiply, regulation becomes further concentrated. What is to constrain FCC's jurisdictional drift? Surely other institutions—state and federal—stand as countervailing forces to telecommunication federalization.

NOTES

1. FCC, In the Matter of Allocation of Frequencies in the Bands Above 890 Megacycles, Docket No. 11866, *Report and Order,* July 29, 1959, 27 FCC 359.

2. FCC, American Telephone and Telegraph Company, Regulations Relating to Connection of Telephone Company Facilities of Customers, Docket No. 12940, *Memorandum, Opinion and Order,* January 17, 1962.

3. FCC, Application of MCI, Incorporated for Construction Permits to Establish New Facilities in the Domestic Public Point to Point Microwave Radio Service in Chicago, Illinois, Saint Louis, Missouri, and Intermediate Points, Docket No. 10509, 1969.

4. FCC, In the Matter of Establishment of Policies and Procedures for Consideration of Application to Provide Specialized Carrier Services in the Domestic Public Point to Point Microwave Radio Service and Proposed Amendments to Parts 21, 43, and 61 of the Commission's Rules, Docket No. 18920, *First Report and Order,* June 3, 1971.

5. FCC, In the Matter of an Inquiry into the Administrative and Regulatory Problems Relating to the Authorization of Commercially Operable Space Communications Systems, Docket No. 14024, *Second Report,* 1961. See also report of the *ad hoc* Carrier Committee, October 12, 1961. FCC, In the Matter of Establishment of Domestic Communication-Satellite Facilities by Non-governmental Entities, Docket No. 16495, *Report and Order,* March 24, 1970.

6. CML Satellite Corporation, 51 FCC 2d (1975).

7. FCC, An Inquiry into the Administrative and Regulatory Problems Relating to the Authorization of Commercially Operable Space Communications Systems, Docket No. 14024, April 1961.

8. AT&T, *Events in Telephone History* (New York: AT&T, 1971), p. 40: "On April 13, 1951, FCC approved Bell acquisition of Western Union telephone business in Pacific Area and sale by Pacific Tel. & Tel. of the telegram business to Western Union." See also Space Satellite Communications, Hearings, Senate Subcommittee on Small Business, 87th Cong., 1st sess. (1961).

9. FCC, Report of the Telephone and Telegraph Committee of the FCC; In the Matter of the Domestic Telegraph Investigation, Docket No. 14650, April 1966.

10. M. Irwin, "Computer Utility: Competition or Monopoly," *Yale Law Journal* 76 (June 1967): 1299.

11. FCC, In the Matter of Regulatory and Policy Problems Presented by the Interdependence of Computer and Communication Services and Facilities, Docket No. 16979, *Final Decision and Order,* March 1971. See also *Tentative Decision,* April 3, 1970, 28 FCC 2d, p. 304.

12. M. R. Irwin, "The Computer Utility: Market Entry in Search of Public Policy," *Journal of Industrial Economics* 27, No. 3 (July 1969): 249.

13. FCC, Computer II Inquiry, Tentative Decision and Further Notice of Inquiry and Rulemaking, vol. 45, no. 20, July 2, 1979; see also *Telecommunication Reports* (May 21, 1979): 1.

14. "AT&T Dataspeed 40 Tariff Stayed," *Electronic News,* February 2, 1976, p. 1.

15. Ibid.

16. FCC, In the Matter of the Application of Satellite Business Systems, File No. 7–DSS–D–76; Reply of ITT, September 24, 1976.

17. 537 F. 2d 787 (4th Cir. 1976) Cert. Denied.

18. IBM v. FCC No. 77–4405, Second Circuit, January 4, 1977; see also AT&T Revisions to Tariff FCC Nos. 260 and 267 Relating to Dataspeed 40 (Dataspeed 40/4), 62 FCC 2d 21 (1977), Transmittal No. 12449, p. 13.

19. FCC, American Telephone and Telegraph Company and the Bell System Operating Companies

Tariff FCC No. 8 (BSoc8), Exchange Network Facilities for Interstate Access (ENFIA), Docket No. 78071, *Memorandum, Opinion, and Order,* April 1982.

20. FCC, Regulatory Policies Concerning Resale and Shared Use of Common Carrier Services and Facilities, Docket No. 20097, *Report and Order,* 1976. See also 60 FCC 2d 261 (1976). Comment: Interdependence of Communications and Data Processing; An Alternative Proposal for the Second Computer Inquiry, *Northwestern Law Review* 73, No. 2 (1978): 318: "If anything the resale and shared use rules appear to broaden the scope of hybrid communications by imposing common carrier status on offers of service with substantial data processing components."

21. Comments of Telenet Communications Corporation, FCC, In the Matter of Regulatory Policies Concerning Resale and Shared Use of Common Carrier Services and Facilities, Docket No. 20097, December 11, 1974, pp. 30–32. See also GTE Service Corporation v. FCC, 474, F2d, 724, 1973.

22. FCC, Amendment of Section 64–702 of the Commission's Rules and Regulations (Second Computer Inquiry), Tentative Decision and Further Notice of Inquiry and Rule Making, Docket No. 20828, 1979, p. 70; *Final Decision,* 1980.

23. FCC, In the Matter of Request for Declaratory Ruling and Investigation by Graphnet Systems, Inc., concerning a proposed offering of Electronic Computer Mail, ECOM, Docket No. 76–6, February 2, 1979. See also *Telecommunication Reports,* No. 13 (April 2, 1979): 16.

24. Proposed Report, Telephone Investigation, Pursuant to Public Resolution No. 8, 74th Cong., 1st sess. (Washington, D.C.: U.S. Government Printing Office, 1938).

25. FCC, Report on the Investigation of the Telephone Industry in the United States, H. R. Doc. No. 340, 76th Cong., 1st sess., 1939.

26. Docket No. 19129, *Final Decision,* 1977.

27. United States v. American Telephone and Telegraph Company, et al., Civil Action No. 74–1698, U.S. District Court, District of Columbia, Brief of the Federal Communications Commission as Amicus Curiae, April 20, 1982, p. 22. See also FCC Supplemental Memorandum 62 FCC 2d 1102: "We urge the court to take no action that will substantially alter the industry structure without initial consideration by the FCC of how such action will comport with the 'public interest.' "

28. FCC, In the Matter of MCI Telecommunications Corporation, Investigation into the Lawfulness of Tariff FCC No. 1 insofar as it purports to offer Execunet Service, Docket No. 2064a, *Decision,* July 13, 1976.

29. MCI Telecommunications Corporation et al., v. FCC et al., Motion for an Order Directing Compliance with Mandate, U.S. Court of Appeals, District of Columbia, April 1978. See also FCC, In the Matter of MTS and WATS Market Structure, Notice of Inquiry and Proposed Rulemaking, March 1978; and MCI Telecommunications Corporation, Microwave Communications Incorporated and N-Triple-6 Incorporated v. FCC, et al., Petition for Review Orders of the FCC, U.S. Court of Appeals, District of Columbia, July 28, 1977, p. 33: "In that eventuality the commission must be ever mindful that just as it is not free to create competition for the sake of competition, it is not free to propagate monopoly for monopoly's sake," (No. 75–1635). See Walter R. Hinchman, Remarks Before the International Communications Association (ICA), May 15, 1978, Las Vegas, Nevada: "By its recent decision the Court of Appeals seems to have taken on the role and function of the FCC as policy maker rather than as a reviewer of the legality of our actions."

30. "A Whole New Way to Figure AT&T's Rates," *Business Week,* February 14, 1977, p. 86: "It's as if we took a journey to the center of the earth and back," says D. A. Irwin. See also R. W. Crandall, "The Impossibility of Regulating Competition in Interstate Markets," Paper delivered at Eastern Economic Association, Boston, May 12, 1979.

31. General Accounting Office, "Legislative and Regulatory Actions Needed to Deal with a Changing Domestic Telecommunications Industry," Washington, D.C., October 1981, Chapter 4.

32. Ibid., p. 65.

33. General Accounting Office, "Can the Federal Communications Commission Successfully Implement Its Computer II Decision?," Washington, D.C., January 29, 1982, p. 5.

34. FCC, Docket No. 19129, *Final Decision,* p. 48.

35. Docket No. 19129, *Initial Decision of Administrative Law Judge,* p. 132. See also FCC, In

the Matter of AT&T, the Associated Bell System Companies, Charges for Interstate Telephone Service, AT&T Transmittal Nos. 10989, 11027, 11650, Docket No. 19129 (Phase II), August 2, 1976.

36. Ibid., p. 48.

37. FCC, In the Matter of Bell System Procurement Practices, Comments of International Telephone and Telegraph Corporation, Docket No. 80–53, October 15, 1981, p. 5; see also C.E. Mayer, "Ex-Employee Urges Breakup of Bell System," *Washington Star*, April 27, 1981, p. F4.

38. Ibid., p. 8; see also F. Barbetta, "ITT Asks FCC to Act on Bell Hardware Procurement Policies," *Electronic News*, June 4, 1979, p. 1; M. J. Richter, "AT&T Rips. ITT 'quota' Buy Proposal," *Electronic News*, July 23, 1979, p. 1.

39. E. Meadows, "Japan Runs into America, Inc.," *Fortune*, March 22, 1982, p. 57.

40. "Why Ma Bell Won't Let a Western Contract Go East," *The Economist*, November 7, 1981, p. 83. See "Japan's High-Tech Sales Hit a U.S. Snag," *Business Week*, January 25, 1982, p. 40; FCC, In the Matter of American Telephone and Telegraph Company and Associated Bell System Operating Companies for authority under Section 214 of the Communication Act of 1934 as amended, to supplement existing facilities by construction, acquisition, and operation of a light guide cable between cities on a main route between New York, New York and Cambridge, MA and between Moseley, Virginia and Washington, D.C., File Ac. WDC 30 71, *Petition to Deny*, Fujitsu America, Inc., December 11, 1981 (hereafter cited as Fiber Optic Docket).

41. Fiber Optic Docket, *Memorandum, Opinion and Authorization*, May 14, 1982, p. 5; see "Legislators Protected Bell-Fujitsu Pact," *Electronic News*, December 7, 1981, p. 21.

42. *Telecommunications Reports* 48, No. 18 (May 3, 1982): 17.

43. FCC, In the Matter of MTS and WATS Market Structure, Docket No. 78-72, Phase I, *Third Report and Order*, February 28, 1983, pp. 124-37.

44. Ibid., Appendix p. 4.

45. John Meyer, Robert Wilson, M. Alan Baughcum, Ellen Burton, and Louis Caouette, *The Economics of Competition in the Telecommunications Industry* (Boston: Charles River Associates, 1979), p. 281.

46. Ian M. Ross, "New Technologies for the Exchange," *Bell Labs Record*, April 1982, p. 83.

47. John Elliot, "The Cult of the Gifted Amateur," *Financial Times*, May 5, 1983, p. 14.

48. FCC, In the Matter of Presten Trucking Company, On Reconsideration of Grant Application for Microwave Facilities in the Motor Carrier Radio Service Inquiry into Certain Arrangements for Cooperative Use of Private Microwave, Docket No. 19309, *Memorandum, Opinion and Order*, August 24, 1971.

49. "Tymeshare Applies for Value-Added Network," *Datamation* (June 1976): 138; also R. Frank, "Tymnet: Packet Switched Net in Disguise," *Computerworld*, October 22, 1974, p. 21.

11

THE COUNTERVAILING FORCES

If federal regulation is frustrated in its ability to deliver even by its own terms, one might speculate that an agency would concede it was overextended. Quite the opposite has occurred, however. The FCC has never evaluated its own economic performance and has failed to identify joint telephone costs. Moreover, it acknowledges that it cannot establish a consensus on carrier pricing strategies, it is frustrated in monitoring cross-subsidization, it is apparently ill-equipped to post demand projections, and it is thwarted in evaluating the economic performance of its constituents. Even so, commission jurisdiction continues to expand and grow.

Despite an explosion in technological alternatives and receding industry boundary lines, the FCC continues to resurrect product, service, and geographic demarcations; to divide, separate, and cartelize markets; and to impose retribution on firms that jump their assigned turf.

And just where precisely is the commission going with its "public interest" mandate? No one, of course, knows for certain. But perhaps its jurisdiction over "enhanced services" is suggestive. The commission maintains that, in permitting AT&T to diversify into information services and products, it will monitor the boundary between regulation and competition. Stated differently, a public utility body will define the demarcation between monopoly and competition. Commission supervision will ensure that monopoly services do not subsidize competitive services.

What is to confine and limit the FCC's jurisdiction to telecommunications? What is to prevent FCC jurisdiction and regulatory control over information, its distribution, use, and content? Surely the Department of Justice, the President, the Department of Commerce, Congress, and the courts stand as countervailing forces against regulatory drift. Let us examine the record of that countervailing power.

DEPARTMENT OF JUSTICE

A survey of the past twenty years of dockets before the FCC reveals that the Department of Justice has been an active participant, filing briefs on customer ownership of terminals, specialized carriers, satellite carriers, computer inquiries, and so on. Yet, from the FCC's vantage point the Department of Justice simply represents one petition among many—essentially a special pleading.

The FCC may elect to ignore the filings from the Department of Justice. Indeed, if it so chooses, the commission may set aside a Consent Judgment entered into by AT&T and the Department of Justice. As noted earlier, a 1949 Justice complaint against AT&T was settled in 1956 by a Consent Judgment. The judgment confined AT&T to regulated communications activities in return for Bell's retention of Western Electric. Obviously, the FCC was neither a party nor a signatory to the 1956 agreement.

Yet, in 1980, the FCC announced it would vacate the substance of that court-sanctioned decree. The commission, reaching for what it termed its ancillary jurisdiction apropos the "public interest" standard, announced that it would permit Bell to form a separate subsidiary as a vehicle to diversify into unregulated information services.[1]

But this jurisdictional reach marked a first step only. In 1982, AT&T and the Department of Justice announced a settlement of their 1974 suit. As discussed earlier, the agreement required Bell to divest its twenty-two operating companies. That same year the FCC stated that, while such a judgment might appear appropriate to the two parties, the commission exercised the final word. Through its control over AT&T's radio licenses, the FCC concluded that it would hold hearings as to whether a transfer of license conflicted with its determination of the public interest.[2]

NATIONAL TELECOMMUNICATIONS INFORMATION AGENCY

If the Department of Justice's effectiveness as a countervailing force is illusive, certainly the White House—the Executive Office of the President—can mold, influence, and exercise control over U.S. telecommunications policy. Perhaps it can, but the record suggests that dealing with an "independent" regulatory agency largely consigns the presidency to a position of observer.

On occasion, the White House will institute a task force to investigate the actions, policies, and directions of the FCC. Such a study may embody a series of prescriptions for future policy consideration. But that is the rub—consideration. The FCC obligingly nods in the direction of such proposals and then proceeds with its own notion of the public interest. In other words, a White House report possesses little clout.

It can be depressing to read White House legal briefs submitted to FCC dockets, and then view the commission conclude that its notion of national telecommunications policy supersedes that of the executive branch of government. Nowhere is that supersession more poignant than in the FCC's decision to extend regulation to firms that lease telephone circuits and then resell them on the open market. White House briefs argued that resellers should be free from common-carrier regulation, free to develop new markets, to engage in price competition, and to let the consumer decide. The FCC replied that a firm buying low and selling high might deem it appropriate to generate a return on its investment. That was enough for the commission. Regulation was duly extended to new entrants over the helpless opposition of the Office of the President.[3]

The White House cannot act as a countervailing force over the FCC because the agency is "independent." The Executive Office is vested with responsibility for U.S. telecommunications policy but lacks the authority to lead or execute. Little wonder that such a policy of disparity invokes ridicule and cynicism. In sum, the Executive Office is ineffective in countervailing FCC jurisdictional creep.

CONGRESS

Can and does Congress pose as a countervailing force against FCC jurisdictional creep? After all, the FCC is a creature of Congress; the agency is beholden to Congress. Congressional action is not inspiring, however. The 1934 Communications Act failed to establish a benchmark to measure the FCC's economic performance. Nor is there a provision that requires Congress to monitor the commission's impact upon economic R&D, efficiency, productivity, innovation, cost, price, investment, and exports. Even if one sweeps such variables under the public interest rubric, Congress has rarely examined whether commission policies enhance or negate economic performance in the United States.

One could put the case negatively. Has Congress assessed the FCC's opportunity cost imposed upon our economy—the cost of efficiency throttled, innovation denied, productivity arrested, consumer choice constrained, and exports stunted? The answer is that Congress has been unable or unwilling to establish economic criteria to assess the FCC's performance record. To that extent, the commission erects its own benchmark and acts as its own arbiter of performance. We may convert the "public interest" into a cliché and say that the FCC in effect judges itself.

Yet, Congress has attempted to impose standards on the commission. In the 1930s, for example, Congress asked the commission to investigate whether telephone companies should buy their equipment on the open market.[4] After several years of investigation, the FCC waffled on the issue. Finally, the commission argued for maintaining the status quo of an equipment cartel.

In the 1960s, after months of strenuous, often vitriolic debate over the nation's satellite policy, Congress imposed a competitive bid requirement upon COMSAT's purchase of related satellite equipment, services, and apparatus.[5] After duly instituting procurement rules, the FCC concluded that competitive bidding by COMSAT was unworkable and/or not feasible. In the meantime, COMSAT acquired an interest in the equipment market. In the instance where Congress did impose rules of conduct and standards of FCC performance, both House and Senate neglected to hold the FCC accountable.

As the FCC writes the nation's strategic scenario in an information-oriented economy, what is the future role of Congress? The answer is that Congress commissions its Office of Technology Assessment (OTA) to study telecommunications technology as it bears on the nation's communications industry. OTA holds seminars and briefings, lets R&D contracts, and issues progress reports. In the meantime, the appropriations for congressional housekeeping expenses grow without interruption.[6]

Since 1976, Congress has held hearings virtually every year, directed toward an inquiry of national communications policy. And virtually every year no legislation has emerged. If patience is a virtue, then both House and Senate are richly endowed. But even more critically, few in Congress have balanced the performance of regulation against the performance of competitive markets and borderless industries. Can the FCC or any regulatory agency match the economic performance of the competitive marketplace? The question cannot be answered if it is not asked.

Having created a regulatory institution, Congress has failed to assess that agency's performance. As this nation moves toward an information economy, Congress gives every indication of manifesting institutional impotence. Having created an independent regulatory agent, Congress either cannot or will not bring that agency to accountability. Congress has failed to impose standards of performance, and in the exceptional instances when such benchmarks are created, Congress has neglected to monitor their enforcement. As a countervailing force against an institution of its own creation, congressional action is largely inaction.

THE COURTS

Has the judicial system served to countervail the commission's jurisdictional drift? On balance, the evidence is that the judiciary has tended both to sanction the extension of FCC regulation into new services and to elevate FCC jurisdiction over that of state regulatory agencies.

On several occasions the courts have drawn the line on commission jurisdiction. In the early 1970s, an appeals court ruled that the FCC's authority did not extend to monitoring transactions between a telephone carrier and its data processing affiliate.[7] In another instance, an appeals court held that the FCC could not segregate and ban entry into long distance switched telephone service.[8] The decision prompted the chief of the FCC's Common Carrier Bureau to suggest that the court decision "appeared to have destroyed the Commission's carefully phased introduction into this market in favor of a free-for-all."[9]

But these court decisions are the exceptions. As a general rule, the courts tend to uphold the preeminence of FCC power. For example, when the commission accepted a tariff on AT&T's computer terminal, firms in the data processing industry objected that the FCC's jurisdiction penetrated into unregulated markets. The court replied that as an institution it was ill-equipped to "second guess" the expertise of a regulatory agency.[10]

A district court ruled that the FCC's jurisdiction preempted the Department of Justice's 1956 Consent Decree. As the court noted, "Nothing in the pleading or in the judgment even remotely suggests that the FCC be subjected to any confinement or restraint in carrying out its mandate by reason of the provision of the judgment.''[11]

The District of Columbia Circuit Court of Appeals also held that the FCC's jurisdiction superseded the 1956 Consent Decree:

Although the Commission recognized that it could not definitively construe the decree, it expressed its belief that the separate subsidiary requirement set forth in the Computer II Decision constituted sufficient ''public regulation'' of AT&T's offering of CPE and enhanced services to satisfy the demands of the Consent Decree.[12]

In return for AT&T's giving up ownership of its Bell operating companies, the 1956 Consent Decree will be lifted and set aside. AT&T, through a separate subsidiary, will be permitted to diversify into unregulated activities. Those activities may range from cable TV to home computers, electronic funds transfer, data processing, and a wide range of information services. The FCC's decision to permit Bell to diversify into unregulated activities in effect abrogates the 1956 judgment.

The problem of power concentration becomes circular. The FCC extends its authority, and judicial review upholds the commission—deferring to a regulator's decision to regulate. This alleged ''alliance'' between the FCC and the District of Columbia Court of Appeals is so blatant that students of the industry assert that judicial power is overly concentrated in Washington.[13] They argue for a redistribution of court venue and caseload assignment.[14] Whatever the answer, if the court system stands as our ''guardian of liberty,'' apparently liberty is equated with a ''federalization'' of telecommunications. The courts do not countervail federal power; rather, they tend to augment and support it.

In sum, the forces arrayed against the jurisdictional expansion of the FCC are fragmented, weak, and vapid. The policies of Congress are to be charitable— irresolute; the executive branch appears to be simply one more solicitor; the Department of Justice is often preempted; and the courts are reluctant to ''second guess'' regulatory expertise.

Undoubtedly, this record is overly pessimistic; indeed, the FCC's record is neither all black nor all white. The commission has sanctioned deregulation of resale carriers, removed regulation of receive only satellite earth terminals, attempted to deregulate international resellers, encouraged COMSAT to engage in direct competition with the overseas record carriers, and has softened its regulation of the broadcast industry.

On the other hand, the FCC remorselessly divides and separates potentially exciting markets of the future—the cellular mobile radio, for example. After a thirteen-year regulatory delay in which the United States frittered away its lead, the FCC then proceeded to cartelize the top thirty metropolitan markets. The commission allocated one frequency to telephone wire line carriers and then invited radio common carriers to compete for the second frequency authorization.

Rivals will compete for cellular radio franchises, while telephone carriers, assured of 50 percent of the market, enjoy a head start.[15]

Any survey suggests that the countervailing forces at the federal level have proved largely impotent. Presumably, the FCC possesses an exclusive lock on defining the content of the public interest. As one commissioner has observed, "Unless we speak out forcefully, no one else will have the public interest as a primary goal."[16] Perhaps, then, state commissions serve as an antidote to federal concentration of telecommunication regulation.

NOTES

1. FCC, Computer II Inquiry, Tentative Decision and Further Notice of Inquiry and Rulemaking, July 2, 1979.

2. United States v. American Telephone and Telegraph Company, et al. Civil Action No. 74–1698, *On Stipulation and Modification of Final Judgement,* Brief of the Federal Communications Commission as Amicus Curiae, April 20, 1982: "The Commission has considerable experience with AT&T and the telecommunication industry after nearly five decades" (p. 29). Also, "The Commission will conduct such further inquiry as may be warranted in its 214 proceedings."

3. FCC, Regulation Policies Concerning Resale and Shared Use of Common Carrier Services and Facilities, Docket No. 20097, *Report and Order,* 1976; see also (Reply Comments), Docket No. 297, Office of Telecommunication Planning (OTP), Executive Office of the President, February 24, 1975, pp. 17 and 18.

4. FCC, Investigation of the Telephone Industry in the United States, 76th Cong., 1st sess., House Document No. 340, Washington, D.C., 1939.

5. See U.S. President's Task Force on Communication Policy, *Final Report,* December 7, 1968, (U.S. Government Printing Office, 1968).

6. James Reston, "A Bonus for Congress," *New York Times,* December 15, 1982, p. A35.

7. GTE Service Corporation v. FCC, 474, F2d, 724, 1973.

8. FCC, In the Matter of MCI Telecommunications Corporation, Investigation into the Lawfulness of Tariff FCC No. 1 insofar as it purports to offer Execunet Service, Docket No. 2064a, *Decision,* July 13, 1976.

9. W. Hinchman, Remarks Before the ICA, Las Vegas, May 15, 1978, p. 7: "The Recent Decision of the D. C. Court of Appeals appears to have destroyed the Commission's carefully phased introduction of competition into the market, in favor of a free for all, open entry policy for the entire long distance communications business." Also, "The Court of Appeals seems to have taken on the role and function of the FCC as policy maker rather than a review of the legality of our actions."

10. IBM v. FCC No. 77–4405, Second Circuit, January 4, 1977.

11. United States v. Western Electric Inc., AT&T, Civil Action No. 17–49, U.S. District Court, District of New Jersey, September 3, 1981. Vincent P. Biunno, District Judge, U.S. District Court, New Jersey. See also A. Pollack, "Bell Upheld in Ruling on Growth," *New York Times,* September 15, 1981, pp. 19, D27.

12. Computer and Communications Industry Association v. FCC, et al., Civil Action No. 81–1193, U.S. Court of Appeals, District of Columbia, November 12, 1982, p. 45.

13. G. O. Robinson, ed., *Communications for Tomorrow* (New York: Praeger Publishers, 1978), p. 430.

14. Ibid., p. 431.

15. "Antitrust Unit Reviewing Radiophone Pact of AT&T, GTE," *Electronics,* July 14, 1982, p. 69; see also M. J. Richter, "Justice to Probe Cellular Arrangements," *Electronic News,* July 12, 1982, p. 1, and John Dizard, "Gold Rush at the FCC," *Fortune,* July 12, 1982, p. 102.

16. *Telecommunications Reports* 48, No. 10 (March 8, 1982): 31. Statement of FCC Commissioner Anne P. Jones.

12

STATE REGULATION

Can state public utility commissions effectively countervail the FCC's jurisdictional drift? The record indicates that state PUCs have engaged in an heroic effort to challenge regulatory policies at the federal level (Table 12.1). State agencies and the National Association of Regulatory Utility Commissions (NARUC), a nonprofit regulatory consortium, have opposed virtually every major federal telecommunications decision of the past twenty years.

In two celebrated cases, the FCC and NARUC confronted each other in judicial appeal. The first case erupted after the FCC opened the interstate toll market to a market entry. NARUC opposed the FCC's decision on the basis that the commission had not inaugurated competitive hearings and that competition would increase rates within their respective jurisdictions. The courts summarized the position of Washington Utilities and Transportation Commission (WUTC), one of the plaintiffs in the suit:

WUTC contends that by authorizing new carriers to furnish specialized interstate communication services without determining that present carriers are unable to meet the need for this service and without individual determination of economic exclusivity, the Commission's order will result in an increase in the number of carriers competing to provide this interstate service and will decrease the usage existing carriers will make of common telephone facilities for the purpose of providing interstate service. This in turn will require allocation to intrastate service of a larger share of the cost of service and equipment used in providing both intrastate and interstate services and will compel WUTC to raise rates for intrastate service contrary to the interests of Washington telephone users.[1]

NARUC further argued that the FCC failed to issue an environmental impact statement regarding its decision to encourage the market entry of specialized carriers. However, in 1975 the circuit court of appeals ruled that the FCC acted within its "statutory authority."[2]

Similarly, state commissions generally opposed the FCC's decision to permit customers to buy, own, and attach telephone equipment to the telephone network.

Table 12.1
NATIONAL TELECOMMUNICATIONS POLICY VERSUS STATE PUCS

Issue	Position	Reason	
Carterfone	Opposition	Jurisdiction	Preempted
Equipment Certification	Opposition	Jurisdiction	Preempted
MCI	Opposition	Jurisdiction	Preempted
Specialized Carrier	Opposition	Jurisdiction	Preempted
Value-added Carrier	Opposition	Jurisdiction	Preempted
Line Sharing	Opposition	Jurisdiction	Preempted
Cellular Radio	Opposition	Jurisdiction	Preempted
Multipoint Distribution System	Opposition	Jurisdiction	Preempted
Xerox's X-10	Opposition	Jurisdiction	Preempted
Private Radio Inter-Connection	Opposition	Jurisdiction	Preempted
Deregulation of Dominant Carriers	Opposition	Jurisdiction	Preempted
One-way Paging	Opposition	Jurisdiction	Preempted
Competitive Equipment Procurement	Silent	Silent	
AT&T Corporate Divestiture	Opposition	Jurisdiction	Preempted

Source: 1981 Report of the Administrative Director on Litigation, Association of Regulatory Utility Commissioners, 1981.

In 1976, an appeals court ruled that the FCC exercised a jurisdiction over terminal attachment that superseded state regulation or rules.

The court challenge arose because sellers of interconnect equipment requested that the FCC preempt state PUCs from banning, restricting, or regulating customer-provided equipment. North Carolina had proposed to prohibit such interconnection, as had the Attorney General of the state of Nebraska. The appeals court once again ruled the preeminence of federal jurisdiction:

If, as North Carolina is formally proposing and the Attorney General of Nebraska has held to be permissible, state jurisdiction over intrastate communication facilities is exercised in such a way that in practical effect either prohibits customer supplied attachments authorized by tariff FCC #263 or restricts their use contrary to the provisions of that or any other interstate tariff, the Commission will be frustrated in the exercise of plenary jurisdiction over the rendition of interstate and foreign communication services that the act has conferred upon it. The Commission must remain free to determine what terminal equipment can safely and advantageously be interconnected with the interstate communications network and how this shall be done."[3]

Obviously, not every state approved customer equipment attachment, resale carrier, or the sale of inside wiring.[4] The California, New York, and Florida commissions were sometimes the exception. Nevertheless, most state agencies pursued a policy of nonentry, noncompetition, and solid opposition to any movement away from telecommunications' status quo.

State policies produced several effects: a "federalization" of U.S. telecommunications policy, a telecommunications underground economy, and a formidable opportunity cost of services and facilities unavailable to the subscribing public.

Prospective firms, sensing a new market or proposing a unique service, have encountered almost solid hostility from state regulatory commissions. The result has been predictable. Potential competitors to telephone companies turned to the federal government for help, assistance, and solace—specifically, to the FCC, and the FCC in turn tended to adopt the new firms. Over time, the commission's jurisdictional reach expanded while state jurisdiction contracted.

Today, the FCC has broadened its regulatory jurisdiction over attachment, local access, enhanced services, depreciation practices, intrastate toll LATA services, and the cost of basic telephone service (the access charge). Regulatory concentration, "federalization" if you will, is the direct result of state regulatory intransigence.[5]

There is yet another consequence. State commissions have promoted the evolution of an information underground. Firms, users, customers, and suppliers employ subterfuge and semantics to circumvent state regulatory rigidity. New telecommunications activities—banned by state regulatory agencies—labeled their activities as "private," "nonprofit," or "interstate" as a means to escape control and regulation.[6] As communication requirements explode, state PUCs act to inhibit intrastate competitors, information services, line sharing and reselling, two-way cable TV, satellite master antenna systems, and communication switching.[7] Such regulatory disincentives drive innovative and state of the art products, facilities, and services underground. In a strange twist of circumstances, state PUCs have become the patron saint of closet services, closet investment, closet systems.

The problem is that this black telecommunications market tends to be confined or restricted to the affluent only. Only the Fortune 500 possess the financial option of going in-house.[8] Only the large firm can take advantage of new, innovative technology as dedicated systems.[9]

The small subscriber does not enjoy a comparable option. Hence individual business users tend to be locked into existing carrier investments and services. To that extent, a firm's options are restricted, products narrowed, flexibility constrained, costs inflated, and competitive response inhibited. State PUC policies, aided and abetted by the National Association of Regulatory Utility Commissions (NARUC) and the Communication Workers of America (CWA), favor the large, well-endowed, affluent corporation and penalize the small prospective firm, thereby creating a bias toward economic concentration. When reminded of the consequence of their actions, state PUCs assert to the critic that they alone understand the content of universal service.

State PUC policy imposes a formidable opportunity cost upon the economy: the cost of services not rendered, products not provided, and systems not developed. State PUC policy has discouraged venture capital, economic innovation, capital formation, technical efficiency, and economic productivity, all the while constraining user choice in the name of the public interest. More profoundly, state commissions have handicapped and restrained this nation's ability to develop an information infrastructure so critical in a world beset by global rivalry and international competition. That such actions are carried out in the name of the public interest only heightens the policy irony.

STATE PUCS AND THE NEWLY INDEPENDENT BELL OPERATING COMPANIES

Today a new problem confronts public utility policy: AT&T's divestiture of $90 billion of assets, 800,000 employees, and twenty-two operating companies. Can state regulators protect the now independent Bell operating companies, their investment, revenues, and markets? Specifically, can state commissions shield the terminal equipment market of the telephone company? That answer is that a terminal is no longer a Model 500 telephone set. A terminal today is a personal computer, an automatic teller machine, a gas pump that accepts credit cards, a satellite receiving dish, a computer mainframe, a television set, an on-line copier, a work station, or even an intelligent office building.[10] Terminals will hardly return to the quiet life of twenty years ago.

A second question is, can state commissions protect the local telephone company's investment from competition? Today over 200 firms sell toll intercity services, and over 180 sell local area networks that employ broadband and baseband architecture.[11] These local area networks incorporate a diversity of transmission media including coaxial cable, fiber optics, paired wire, and free space optical links. Inasmuch as 60 percent of all data generated never leaves a building, a revolution is brewing within the laboratory, university, office, factory, bank, and store. As microelectronic costs drop and as work stations proliferate, the demand for high-capacity networks can be expected to increase rather than decrease in the 1980s.

Third, can state PUCs narrow the technical options that invite bypass of local telephone plant and facilities? If state commissions exercised discretion over the range of options available to the user or to outside firms, the answer would be perhaps. Consider the following:

- Satellite dishes
- Satellite master antenna television (SMATV)
- Direct broadcast satellites
- Cellular radio

- Infrared data techniques
- Coaxial cable
- Fiber optics
- Digital transmission systems
- FM broadcast subcarrier systems
- Lower power television[12]
- Free space optical communications

And technology now permits combinations and permutations of the above. Note that in Figure 12.1 the link of satellites to local area networks in the United Kingdom portends still another specie of loop bypass.[13] Cellular radio poses an intriguing alternative to paired wire despite the FCC's thirteen-year delay of that market.[14]

Fourth, can state PUCs prohibit customers, firms, and toll carriers from employing bypass techniques? State commissions will not want for trying.[15] The California Public Utility Commission implored the federal judge in the Bell divestiture case to block AT&T from bypassing its former progeny at the local level. The California commission catalogued the potential threat as:

- Large corporate headquarters in a single building or a single office complex
- New industrial or commercial tracts or large office buildings owning their own communication networks within the tract or building and furnishing services to all occupants
- Large scale operations having a high volume of traffic to coordinate with computers in other cities
- Large government office concentrations
- University campuses.[16]

Judge Harold Greene of the district court rejected the California commission's request to ban technological bypass. He observed: "Neither the court nor those who object to the decree can halt the electronic revolution any more than the Luddites could stop the industrial revolution at the beginning of the last century."[17]

Moreover, a new species looms on the horizon, the local loop reseller; witness the New York Port Authority's consortium to erect seventeen satellite earth terminals on Staten Island and to relay voice, data, facsimile, and teleconferencing in and out of Manhattan through fiber optic cable. Merrill Lynch, a member of the consortium, has intimated that it might become a local loop reseller.[18] And even the United Kingdom is stirring; a consortium has acquired the water rights of way under London for purposes of laying a fiber optic cable.[19]

Fifth, can state commissions protect basic telephone rates? The answer is each state will try. Some state commissions will attempt to block entry of long distance companies within their jurisdictions on grounds of revenue creamskimming and

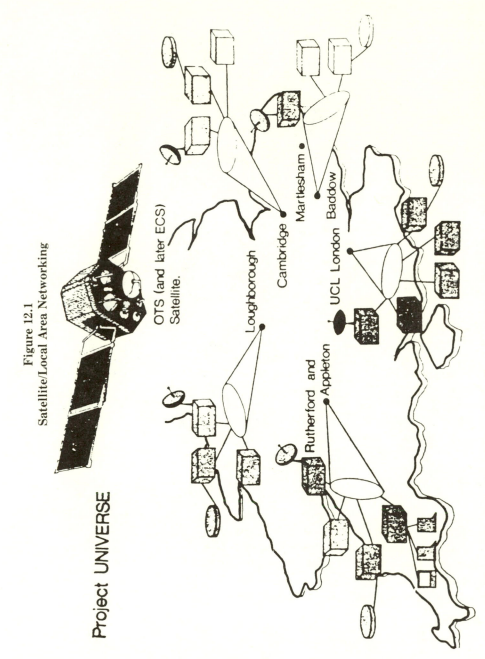

Figure 12.1
Satellite/Local Area Networking

Project UNIVERSE

OTS (and later ECS)
Satellite.

Loughborough

Cambridge

Martlesham

Baddow

UCL London

Rutherford and
Appleton

Source: Elaine Williams, "Project Universe Shrinks the Computer World," *Financial Times,* February 4, 1983, p. 19.

wasteful plant duplication. Other states will attempt to confine access revenues within their individual jurisdictions.

State Commissions assume responsibility for a policy of rate base economics, cost plus pricings, extended depreciation lives, flat telephone tariffs, and a ban on market entry at the exchange level.[20] The same commissions however, see no connection between these economic disincentives and obsolete switching plant, narrow options to inside wire, a paucity of research in loop distribution, a telephone system "strangling in wire" and restricted capacity of copper wire.[21] Nor do State Commissions comprehend the very essence of competition—its dynamic interplay in stimulating the innovation of digital compression techniques, the search for technical options and combinations, and sensitivity to emerging needs of the consuming public. Any short fall in telephone carrier performance must be assigned to the institutionalized disincentive embodied in state utility regulation.

And now state commissions are mounting a legislative endeavor to insulate local telephone companies from technological diversity. Proposed federal legislation either contemplates the outright prohibition of technical bypass of local distribution facilities (lines and switching) or proposes a "tax" on bypass technology in the name of ensuring basic telephone service.

Yet telecommunication technology continues apace. Today "intelligent" office buildings—imbedded with fiber optic cable, microprocessors, and satellite dishes—are under construction throughout the country. Tomorrow, these edifices of shared telecommunication will constitute still another component of an information matrix. Today, condominiums and co-op apartments are embedded with micro-electronics. Tomorrow, these structures will be an increment to a national telecommunications network.

Unintimidated, state public utility commissions are moving to ban, protect, tax, or regulate such an infrastructure into oblivion—all in the name of universal telephone service. Over the past two decades, state regulation and its trade association, NARUC, have viewed every technological alternative as a competitive threat to telephone rates. That opposition, that intransigence, that misperception has "federalized" telecommunication regulation, precipitated an underground information economy, and imposed a formidable opportunity cost of services denied, investment forestalled, choice constrained.

Given the massive restructuring of the Bell System and the phase-down of the toll revenue subsidy to basic exchange rates, will state regulation be driven to oppose the evolution of tomorrow's information infrastructure? If the past is at all indicative, the answer is not reassuring.

NOTES

1. Washington Utilities and Transportation Commission v. FCC, 1513 F.dd 1147, 1975.
2. Ibid., 1148.
3. North Carolina Utility Commission v. FCC, 537, F. dd (1976), p. 793.
4. "Florida Ok's Intrastate Discount Service by Resellers," *Electronic News*, December 20, 1982, p. 43.

5. Computer and Communications Industry Association v. FCC et al., U.S. Court of Appeals, District of Columbia, November 12, 1983, p. 10. See also *Telecommunications Reports*, March 29, 1982, p. 3.

6. Colin Ungaro, "New and Old Players Set Stage for Networking Battles," *Data Communications* (December 1981): 154; "Dialing Locally—Without AT&T," *Business Week*, October 5, 1981, p. 102.

7. "The Resale Business in Phone Lines," *Business Week*, July 13, 1981, p. 68; Dale Hatfield, "Local Distribution: The Next Frontier," paper delivered to the Ninth Annual Telecommunications Policy Research Conference, Annapolis, Maryland, April 29, 1981; "Telecommunication," *Business Week*, October 11, 1982, p. 60.

8. John Meyer, Robert Wilson, M. Alan Baughcum, Ellen Burton, and Louis Caouvette, *The Economics of Competition in the Telecommunications Industry* (Boston, Mass.: Charles River Associates, 1979), p. 281.

9. J. Alleman and E. Beauvais, "No Main Is An Island or Local Loops As Barriers to Entry," paper delivered at the *Western Economic Association Annual Meeting*, July 5, 1981, p. 13; see also Policy Development Division, Federal Communications Commission, "The Future of Digital Technology in the Private Radio Services," August 1981, p. 21.

10. "Gasoline Stations Operate Without Attendants: Motorists Just Insert Credit Cards, Fill 'er Up," *Marketing News*, November 26, 1983, p. 19. See also J. Blyskal, "Don't Hold the Phone," *Forbes*, January 3, 1983, p. 43; and "AT&T to Design Phones for 5-year Life," *Electronics*, December 29, 1982, p. 36.

11. "Linking Up to Boost the Office," *Business Week*, March 21, 1983, p. 144; C. Anne Prescott, "Packaging a 'Genie'," *Bell Telephone Magazine* 5 (1982): 10.

12. Dale Hatfield, "Technological Transformation of Telecommunications and the Public Policy Response," *Institute of Public Utilities Conference*, Williamsburg, Va., December 3, 1980, p. 5.

13. Elaine Williams, "Project Universe Shrinks the Computer World," *Financial Times*, February 4, 1983, p. 19.

14. "Carrier Pigeons Ferrying Lockheed Microfilm," *New York Times*, August 19, 1982, p. 20: "It costs $10.00 a print to use machine (computer linked to computer) but the pigeons cost $1.00."

15. "Justice Raps Ambiguity in Cellular License Order," *Electronic News*, December 6, 1981, p. 1; and S. Crump, "Cellular Radio Is More Than Mobile; It'll Be the Telephone of the Future," *Communication News* 18 (August 1981): 46.

16. U.S. v. American Telephone and Telegraph Company, et. al. Civil Action No. 74–1690, U.S. District Court, District of Columbia, Brief of the People of the State of California and the Public Utilities Commission of the State of California on Issues Regarding the Absence of Restrictions on AT&T, June 13, 1982, p. 7.

17. United States v. American Telephone and Telegraph Co., *Opinion*, Judge Harold Greene, August 11, 1982, p. 70.

18. Paul Betts, "Beaming on Staten Is.," *Financial Times*, December 8, 1982, p. VI.

19. "Telecom Access is Where and When You Find It," *Telecom Times and Trends* 2, No. 1 (January 1983): 3.

20. V. Perry, "Telco Equipment Depreciation Policies: Why Major Changes Are Needed Fast," *Telephony*, September 8, 1980, p. 84; and G. Putka, "AT&T Depreciation Schedule Is a Major Issue," *Wall Street Journal*, January 13, 1983, p. 31.

21. FCC. In the Matter of American Telephone and Telegraph Company, The Associated Bell System Companies, Docket 19129, Phase II, FCC Trial Staff Submission, 1975, Trial Staff Exhibit 275. A 1973 Bell report observes, "there is no real work being done either in the research area or the development area to solve the local plant problem."

13

CONCLUSION

The world of regulation and the world of technology stand in sharp contrast. Regulation is reluctant to acknowledge that research is an extension of unprecedented competition and rivalry. Rather, the regulator tends to view R&D as autonomous and above the petty forces of the marketplace.

The world of regulation is premised upon a single industry firm, boundary lines established by tradition, insulated by geography, promoted by due process. Because a monopoly firm is presumed to be walled by physical location, regulation's mandate is to monitor the firm's performance on behalf of the consumer. Accordingly, the world of regulation sanctions market access to designated suppliers only and then only reluctantly. Oversight continues in the name of the public interest.

Institutionalized regulation distrusts market entry, product diversity, price competition, and individualized requirements. The world of regulation is more congenial to an environment of homogeneity, order, and predictability. The world of regulation abhors surprises and distrusts rivalry.

The world of regulation looks askance at choice. In the eyes of the regulator, consumer options are contrived rather than genuine. Regulation much prefers a narrow range of alternatives endowed with qualities deemed appropriate by commission expertise and approval. Basically, regulation distrusts the consumer's judgment.

Finally, the world of regulation misunderstands dynamic change. Change disrupts the status quo, the premise of development cycles, the longevity of economic life, and the assumption of product depreciation and price elasticity. Technology assaults the mechanics of cost plus pricing. Over the long term, technology acts to subvert regulatory authority and the relevance of quiet due process.

Technology, in contrast to regulation, is funded by dozens of industries as an extension of multi-industry competition. The microelectronics revolution is nothing if not global and international.

Industry boundary lines are eroding and shifting. Information networks evolve within buildings, factories, offices, and laboratories. Regional, national, and international communication links are being forged unrelentingly as information distribution systems penetrate new markets and facilitate new services. The evolution of global markets without boundaries is emerging as a new economic imperative of the 1980s.

Nearly two dozen industries are vectoring toward the same revenues, customers, and services. The ingredients of intensified interindustry competition are erupting in the automation of the factory, office, store, home, school, and bank. Competition is generating specialization and market segmentation. Interindustry overlap can be seen in the following:

- Penny's v. Marriott Inn
- Savings banks competing with Sears Roebuck
- Computer firms competing with telephone carriers
- Newspaper companies competing with cable TV
- Cable TV competing with the broadcast industry
- Data processing firms competing with the U.S. Post Office
- Federal Express competing with satellite carriers
- New York Times competing with telephone companies
- Exxon competing with computer firms
- Sony Corporation competing with IBM
- Sears competing with AT&T
- Bekins Moving Company pitted against Federal Express.
- Holiday Inn v. American Airlines

It is critical to note that consumer options stand as the mirror image of multi-industry entry. The expanding choices in chips, semiconductors, satellites, microwave, software, and hardware enlarge the tradeoffs available to producer and consumer. Consider the abundance of options.

- Video-disc versus video cassette recorders
- Travel versus teleconferencing
- Home video versus classroom lecture
- Videotext versus banking
- Electronic mail versus post office
- Broadcast signal versus cable TV
- Newspapers versus teletext
- Cellular radio versus wire line telephone companies
- Computer services versus tax service
- Videotext versus shopping

- Home security versus personal guard
- On-line bibliographic services versus public library
- Direct broadcast satellites versus commercial satellites
- Facsimile service versus commercial letter
- Fiber optics versus copper cable
- Local area networks versus PBXs
- Satellite master antenna systems versus cable TV.

A renaissance of tradeoffs lies before us.

Finally, the pace of technological change shows little sign of slowing down. Microelectronic costs and prices continue to fall and no end appears to be in sight. The world of the 1980s suggests an environment of intense volatility, expedited innovation, unprecedented rivalry. If nothing else, the world of technology beckons, invites, and solicits the entrepreneur.

A hierarchy of networks—local area, local distribution, regional, international—marks the essence of an emerging information infrastructure. This diversity of intelligent appliances interacting with a widening hierarchy of networks now influences our work, leisure, investment, health, and education. Network proliferation is redefining the firm, the store, the bank, the office, the home, the shop, the classroom. An information infrastructure is beginning to impact resource allocation, plant location, job opportunities, the mix of the nation's economy and exports. A threshold of new services, new products, and unprecedented options heralds an era of abundance and growth.

In this juxtaposition of technology and due process where do our institutions of regulatory oversight reside? The performance of the FCC must be judged against that of another institution—that of the competitive market. Can the FCC deliver in an age of borderless markets, intensified domestic rivalry, and enhanced global competition? We think not. On balance, the commission's record fosters economic concentration, cartelizes markets, waffles on rate structure cases, throttles innovation, protects manufacturing monopolies—all in the name of the public interest. Clearly, the commission has never attempted to assess the economic performance of itself or its clients; and clearly Congress has been unable to render an audit that assesses commission economic performance.

Can state PUC regulation assume the burden of supervision in a world of falling boundaries and intensified competition? The past record of state commissions is singularly uninspiring. State commissions have resisted market entry, protected obsolete plants, embraced cost plus pricing, elongated depreciation lives, curtailed telephone service, restricted consumer options, sanctioned production cartels, and solicited the flow of endless subsidies. State intransigence has concentrated regulation at the federal level and has precipitated the formation of a subterranean information economy. When one balances the record of state regulation against the imperatives of an information infrastructure, it is difficult to resist the conclusion that commission regulation constitutes institutional insolvency.

Nevertheless, state regulation possesses one virtue that supersedes that of federal power. State utility commissions bear the consequences and burden of their regulatory conduct. If state commissions insist that an optimal strategy is to control entry, cap profits, regulate profits, impose economic disincentives, and police electronic funds transfer, electronic mail, voice, graphics, teleconferencing, local area networks, teletext, and view data—all under the presumption of pursuing the public interest—then so be it.[1]

Bank credit card operations can move to Sioux Falls, South Dakota (Citicorp); traveler check operations to Salt Lake City (American Express); credit processing centers to Tampa, Florida (Citicorp); stock and bond operations to Jacksonville, Florida (AT&T); banking/insurance services to Rapid City, South Dakota (Citicorp); computer operations to Princeton, New Jersey (Merrill Lynch) or to Westchester County (Citibank); data processing plants to Plantation, Florida (American Express).[2]

An information infrastructure facilitates the mobility of capital and labor resources. Any state that obstructs this emerging environment will be economically accountable for its tax, regulatory, and risk disincentives. On the other hand, the state that contracts its regulatory reach, promotes research, encourages entrepreneurship, and fosters both investment, risk and consumer options, holds the promise of comporting with the competitive imperatives of an information economy.

In a post-industrial economy each state must search for its own identity, strategy, and philosophy. Such is our legacy of pluralism and diversity. The issue, therefore, is not technology versus regulation. Either that matter is perceived and acknowledged or it is not. Rather, the question is who should make critical information policy—the federal government or state government? We assert each state stands supreme.

In sum, Theodore Vail bequeathed two institutions to the United States: telephone monopoly and commission regulation. Technology, time, and circumstance have dissipated the former. Whether in the United States or Canada, telephone carriers are moving away from an environment of sheltered regulation in order to seek and embrace the opportunities of an information age.[3] Incrementally telephone carriers are beginning to vacate the public utility principle.

But what of Vail's second legacy, commission oversight? It would be one thing to preserve that institution if its actions were trivial and benign. But when regulation positions itself against an emerging information infrastructure, then the consequences of that jurisdictional reach are central and overriding.

Private institutions experience an appropriate rise and fall. That is to be expected. But are public institutions destined to live in perpetuity?

NOTES

1. "Florida OK's Intrastate Discount Service by Resellers," *Electronic News,* December 20, 1982, p. 43; "Pacific Telephone Asks California to Restrict Long Distance Service," *New York Times,* April 29, 1983, p. 48.

2. David Bird, "Citicorp to Shift 600 Jobs in City to Tampa Base," *New York Times,* April 27, 1983, p. B1. Andrew Pollack, "Role of Telecommunications in Industrial Planning Grows," *New York Times,* May 2, 1983, p. 1; Robert A. Bennett, "Citibank Move into Insurance," *New York Times,* March 30, 1983, p. D1; John Herbers, "Flight of New Industries Worries Minnesotans," *New York Times,* April 12, 1983, p. D1; James Barron, "Long Island is Winning a Profitable Paper Chase," *New York Times,* February 20, 1983, p. E9; Julie Salamon, "Citicorp to Buy Control of Bank in South Dakota," *Wall Street Journal,* March 30, 1983, p. 5; Randall Smith, "Some Unlikely Places Benefit from the Boom in Financial Services," *Wall Street Journal,* March 31, 1983, p. 1; Tim Carrington and Daniel Hertzberg, "American Express Tries a Difficult Game in Its Expansion Drive," *Wall Street Journal,* February 25, 1982, p. 1; Timothy D. Schellhardt, "War Among the States for Jobs and Business Becomes ever Fiercer," *Wall Street Journal,* February 14, 1983, p. 1.

3. Restrictive Trade Practices Commission, *Telecommunications in Canada,* Part II, The Proposed Reorganization of Bell Canada, Consumer and Corporate Affairs, Ottawa, Canada, 1982; also *The Wall Street Journal,* "Bell Canada Reorganizes Itself," June 24, 1982, p. 8.

BIBLIOGRAPHY

PERIODICALS

Alexander, T. "The Postal Service Would Like To Be the Electronic Mailman Too." *Fortune*, Vol. 102, June 18, 1978, pp. 92–93.

Anders, C., and Dodush, M. "Baldwin-United in Reversal, Plans to Use Bank Credit in Financing MGK Take-Over." *Wall Street Journal*, February 9, 1982, p. 10.

"ANPA Files Motion to Enter AT&T Case." *Editor & Publisher*, Vol. 114, April 4, 1981, p. 12.

"Antitrust Unit Reviewing Radiophone Pack of AT&T, GTE." *Electronics*, Vol. 55, July 14, 1982, p. 69.

Armington, C., and Odle, M. "Small Business—How Many Jobs?" *The Brookings Review*, Vol. 1, No. 2, Winter 1982, pp. 14–17.

Arnold, W., and Rausch, J. "Teleconferencing vs. $21 Billion Annual U.S. Business Travel." *Electronic Business*, Vol. 10, June 1981, p. 76.

"AT&T Cancels Austin Test of Electronic Yellow Pages." *Editor & Publisher*, Vol. 114, July 18, 1981, p. 7.

"AT&T Dataspeed 40 Tariff Stayed." *Electronic News*, Vol. 22, February 22, 1976, p. 1.

AT&T Says It Will Look Abroad to Offer Same Debt and Build Production Plants." *Wall Street Journal*, January 20, 1982, p. 8.

"AT&T Settles Datran Suit for $50M." *Electronic News*, Vol. 1, March 17, 1981, p. 1.

"AT&T Slammed on EIS II (Electronic Information System II)." *Datamation*, Vol. 27, July 1981, p. 61.

"AT&T to Design Phones for 5 Year Life." *Electronics*, December 29, 1982, p. 36.

Atsuyoshi, O; Fakemochi, I; and Masanori, M. "Electronic Shock: The Impact of Microelectronics." *Japan Echo*, Vol. 9, November 1982, pp. 18–35.

Aug, S. "AT&T Puts Cost of U.S. Antitrust at $1 Billion." *Washington Star*, February 17, 1978, p. 20.

Barbetta, F. "AT&T Marketing Realignment Plan." *Electronic News*, Vol. 24, February 27, 1978, p. 1.

————. "ITT Asks FCC to Act on Bell Hardware Procurement Policies." *Electronic News*, Vol. 25, June 4, 1979, p. 1.

Barna, B. "McAuto on the Move." *Datamation*, Vol. 27, May 1981, p. 75.

Batt, R. "Vitalink Enters Satellite Earth Station Area." *Computerworld*, Vol. 15, November 30, 1981, p. 7.

Baur, F. "Microelectronic in International Competition." *Siemens Review*, Vol. 9, No. 6, 1980, p. 4.

"Bekins: A Household Moving Leader Highballing Into Related Services." *Business Week*, November 1, 1982, pp. 58–59.

"Bell Asks Teleconferencing Service OK." *Electronic News*, Vol. 27, March 23, 1981, p. 20.

"Bell Labs Answers Calls for Help." *Business Week*, January 23, 1981, p. 38.

Bennett, R. A. "A Bank By Any Other Name." *New York Times*, December 27, 1981, Sec. 3, p. 1(F).

————. "Citicorp's Satellite Challenge." *New York Times*, March 24, 1983, p. D1.

————. "Wriston Ponders Bankless Citicorp." *New York Times*, July 31, 1981, p. D1.

Bergreen, L. "Hello LA? Miami? Hong Kong?" *New York Times*, August 30, 1981, p. 5.

Betts, Paul. "Baby Bell Starts Work on a New Image." *Financial Times*, January 20, 1983, p. 7.

————. "Beaming on Staten Island." *Financial Times*, December 8, 1982, p. V1.

————. "How Ma Bell Plans to Ring the Charge." *Financial Times*, April 1, 1982.

————. "U.S. Telephone Service: Plain and Simple It Won't Be." *Financial Times*, January 12, 1983, p. 12.

Birch, D. "Who Creates Jobs." *The Public Interest*, No. 65, Fall 1981, pp. 3–14.

Blyskal, J. "Don't Hold the Phone." *Forbes*, Vol. 131, January 3, 1983, p. 43.

Branscomb, L. M. "Computing and Communications—A Perspective of the Evolving Environment." *IBM Systems Journal*, Vol. 18, No. 2, 1979, pp. 189–201.

Briggs, J. A. "The Pace Setter." *Forbes*, January 3, 1983, pp. 42–43.

Brinton, J. "R&D: Some Strategies that Work." *Electronics*, Vol. 53, April 24, 1980, p. 94.

Brovsell, D. "Citicorp Effort to Provide DP Services Gains in Court." *Electronic News*, Vol. 28, April 5, 1982, p. 1.

Brown, W. "Western Electric Modernization Means Pain for Baltimore." *Washington Post*, January 31, 1983, p. 17.

Bylinski, G. "The Game Has Changed in Big Computers." *Fortune*, Vol. 105, January 25, 1982, p. 82.

"CAD Redraws the Architects' Job." *Business Week*, March 15, 1982, pp. 134–36.

Cane, A. "Why IBM Will Shake the Videotext Game." *Financial Times*, November 2, 1983, p. 14.

"Carrier Pigeons Ferrying Lockheed Microfilm." *New York Times*, August 19, 1982, p. 20.

"Catalogue Cornucopia." *Time*, November 8, 1982, p. 71.

"CBS is Cleared to Enter Cable TV Market: FCC May Consider Repeat of General Ban." *Wall Street Journal*, August 5, 1981, p. 6.

Cellis, R. "A Bid to Revitalize U.K. Small Business Sector." *International Herald Tribune*, October 27, 1982, p. 9S.

————. "Risk Takers Find Terms Favor the Conservatives." *International Herald Tribune*, October 27, 1982, p. 8S.

Charlish, C. "EGB Moves Message with Power Through Grid." *Financial Times*, September 27, 1982, p. 12.

"Citicorp Steals a Piece of the Action." *The Economist*, Vol. 278, February 7, 1981, p. 75.

"ComSat Buys IC Firm." *Electronics*, Vol. 55, February 10, 1982, p. 63.

"Communications Dogfight." *Dun's Review*, Vol. 12, June 1977, p. 48.

"Communications: Who Will Supply the Office of the Future." *Business Week*, July 27, 1974, pp. 42–50.

Cooney, S. "Lowering Skies for the Satellite Business." *Fortune*, Vol. 106, December 13, 1982, p. 148.

"Crossed Wires for American Information Network." *New Scientist*, Vol. 84, December 6, 1979, p. 778.

Crump, S. "Cellular Radio is More than Mobile: It'll be the Telephone of the Future." *Communication News*, Vol. 18, No. 8, August 1981, p. 46.

"Datran Settlement." *Datamation*, Vol. 26, April 1980, p. 88.

"A David-Goliath Threat to Cable." *Business Week*, August 16, 1982, p. 106.

de Jonquieres, Guy. "More Room in Ma Bell's Auberge." *Europe*, No. 227, September-October 1981, pp. 36–38.

————. "Wang: Relying on Nimble Feet to Break Into the Big League." *Financial Times*, March 18, 1981, p. 14.

————. "We'll Talk to Anyone Who Has a Better Mousetrap." *Financial Times*, January 24, 1983, p. 12.

"Dialing Locally—Without AT&T." *Business Week*, October 5, 1981, p. 102.

Dizard, John. "Gold Rush at the FCC." *Fortune*, Vol. 106, July 12, 1982, p. 102.

Dooley, Ann. "Union Leader Sets Sights on the Automated Office." *Computerworld*, Vol. 15, July 27, 1981, p. 1.

"Dow Jones to Buy Part of W. V.'s Westar V." *Communication News*, Vol. 18, No. 3, March 1981, p. 8.

Drayfack, K. "France Wants a Bigger Piece of Pie." *Electronics*, Vol. 54, October 23, 1981, p. 98.

Egan, M. "Motorola Revamping Aims at Data Com Net." *Management Information Systems Week*, Vol. 13, January 22, 1982, p. 1.

Eger, John. "Bell's End Run." *Datamation*, Vol. 23, May 1977, pp. 81–82.

Ela, J., and Irwin, M. R. "Blindside Competition, Technological Change Demand Different Strategies in Information Age." *Marketing News*, Vol. 16, no. 11, November 26, 1982, p. 1.

"The Electronic Office is Temporarily on Hold." *The Economist*, Vol. 275, June 14, 1980, pp. 71–72.

Elion, John. "The Cult of the Gifted Amateur." *The Financial Times*, May 5, 1983, p. 14.

Emmett, R. "Return of the Vikings." *Datamation*, Vol. 27, no. 4, April 1981, p. 83.

Ergas, H. "Should There Be More Competition in Telecommunications?" *OECD Observor*, No. 121, March 1983, pp. 30–33.

"Ericsson Moves on the U.S. Giants." *Business Week*, March 22, 1982, p. 74.

Feldman, R. "AT&T Reported Urging Users to ByPass Local Loops." *Management Information Systems Week*, Vol. 3, December 22, 1982, p. 14.

Fields, S. "Intel Puts More Chips on Computers." *Electronics,* Vol. 54, No. 2, February 10, 1982.

"Final Shape of Big AT&T Settlement." *U.S. News and World Report,* Vol. 93, No. 9, August 30, 1982, p. 36.

"Five Ways to go Bust." *The Economist,* Vol. 286, January 8, 1983, p. 11.

"Florida OK's Intrastate Discount Service by Resellers." *Electronic News,* Vol. 28, No. 1425, December 20, 1982, p. 43.

"France is Disconnected in an AT&T Philips Link." *Business Week,* October 11, 1982, p. 47.

Frank, R. "A Big Blow to AT&T." *Datamation,* Vol. 26, August 1980, p. 57.

———. "Citibank Reported Running Private Packet Net." *Computerworld,* Vol. 11, November 14, 1977, p. 53.

———. "IBM, CBEMA Tell FCC not to Tariff Bell Dataspeed 40." *Computerworld,* Vol. 10, January 12, 1976, p. 23.

———. "Tymnet: Packet Switched Net in Disguise." *Computerworld,* Vol. 8, October 22, 1974, p. 21.

Frasee, J. "AT&T's deButtes Blasts Interconnects." *Electronic News,* Vol. 18, October 16, 1972, p. 17.

Frey, Annette. "The Public Must Be Served." *Bell Telephone Magazine,* No. 59, March/June 1975, p. 5.

"Fujitsu Protests WE Pact Award." *Electronic News,* Vol. 27, No. 16, November 16, 1981, p. 1.

Furst, A. "A Fresh Crop of Fail-Safe Computer." *Electronic Business,* Vol. 7, No. 10, October 1981, p. 58.

Gamble, Richard. "The Early Competitive Era in Telephone Communications, 1893–1920." *Law and Contemporary Problems,* Vol. 34, Spring 1969, pp. 347–50.

"Gannett's National Gamble." *Newsweek,* Vol. 100, No. 12, September 20, 1982, p. 101.

"Gasoline Stations Operate without Attendants, Motorists Just Insert Credit Cards, Fill 'er Up." *Marketing News,* Vol. 16, No. 11, November 26, 1983, p. 19.

Gigot, P., and Lueck, T. "Sears Expansion Brings Increased Competition to Banks and Brokers." *Wall Street Journal,* October 12, 1981, p. 1.

Grace, S. "Rising Rivalry, AT&T and IBM Tread on Each Other's Toes, as Courses Converge." *Wall Street Journal,* September 4, 1981, p. 1.

Hardt, T. S. "Thrift Cleared to Enter Securities Field by Bank Board; Legal Challenges Likely." *Wall Street Journal,* April 4, 1982, p. 21.

"Harvester Cleared for Phone Service." *New York Times,* April 10, 1982, p. 30.

Hindin, H. "Large Scale Integration Latches on to the Phone System." *Electronics,* Vol. 53, June 5, 1980, pp. 113–27.

Hirsch, P. "AT&T's Big Plans for a Value Added Service." *Datamation,* Vol. 22, January 1976, pp. 100–102.

———. "Postal Commission to Suspend ECOM Inquiry." *Computerworld,* Vol. 15, December 14, 1981.

———. "Tymeshare Plans to Acquire Microband Corp. of America." *Computerworld,* Vol. 15, September 1, 1981.

Holmes, E. "FCC Defines 'DP' as Difference Between Smart, Dumb Terminals." *Computerworld,* Vol. 9, February 28, 1977, p. 6.

Holsendolph, E. "Key to Legal Status of Telephone Accord." *New York Times*, January 14, 1980, p. 1.

————. "MCI to go Global by Buying WUI." *New York Times*, December 16, 1981, p. D5.

Irwin, M. "The Computer Utility: Market Entry In Search of Public Policy." *Journal of Industrial Economics*, Vol. 17, No. 3, July 1969, pp. 239–52.

————. "Telecom User Pressure Is Changing Market Boundaries." *Telecommunications*, Vol. 16, No. 100, September 1982, pp. 17–20.

————, and Ela, J. "Commission Regulations under Technological Stress: The Case of Information Services." *Public Utilities Fortnightly*, Vol. 107, No. 3, June 18, 1981, pp. 1–8.

————, and Ela, J. "U.S. Telecommunications Regulation: Will Technology Decide." *Telecommunications Policy*, Vol. 5, No. 1, March 1981, pp. 24–32.

————, and Janisch, H. "Information Technology and Public Policy: Regulatory Implications for Canada." *Osgoode Hall Law Journal*, Vol. 20, No. 3, September 1982, pp. 610–41.

————, and McConnaughey, James. "Rate Base Economies and Vertical Integration: Shifting Standards in Telephone Regulation." *Indiana Law Journal*, Vol. 54, No. 2, 1979, pp. 185–200.

————. "Local Networks In the U.S.: Technology vs. Regulation." *Telecommunications*, Vol. 16, No. 3, March 1982, pp. 17–21.

"Is That You, Ma Bell?" *Sales Management*, Vol. 115, No. 8, March 3, 1975, pp. 30–34.

"The Japanese Invasion: Chips Now, Computers Next?" *Electronic Business*, Vol. 10, July 1981, pp. 84–87.

Johnes, K. "Texas Publishing Execs Hail 7-Year News Ban on AT&T." *Management Information Systems Week*, Vol. 13, September 1, 1982, p. 8.

Jordan, L. "Teleconferencing: Friend or Foe." *Lodging*, Vol. 7, No. 4, January 1982, p. 37.

"Justice Raps Ambiguity in Cellular License Order." *Electronic News*, Vol. 27, December 6, 1981, p. 1.

Killingsworth, V. "Corporate Star Wars: AT&T vs. IBM." *Atlantic*, Vol. 243, No. 5, May 1979, pp. 68–75.

Kindel, S., and Schriber, J. "The Old Boy Satellite Network." *Forbes*, Vol. 13, November 8, 1982, pp. 113–14.

Kleinfield, N. "Newspapers Stalk Cable TV." *New York Times*, June 25, 1981, pp. D1–D17.

Knox, M. "PBX Life Cycle Called Key to Telecom Industry Future." *Management Information Systems Week*, Vol. 12, August 19, 1982, p. 11.

Kobayashi, K. "Telecommunications and Computers: An Inevitable Marriage." *Telephony*, Vol. 198, No. 4, January 23, 1980, pp. 78–86.

"Ky Food Marts Chosen to Provide EFT Services." *Management Information Systems Week*, Vol. 13, February 3, 1982, p. 12.

Langdon, W. C. "The Myths of Telephone History." *Bell Telephone Quarterly*, Vol. 13, No. 2, April 1933, pp. 123–40.

"Legislators Protest Bell—Fujitsu Pact." *Electronic News*, Vol. 27, December 7, 1981, p. 21.

Lehner, C. "Technology Duel: Japan Strives to Move from Fine Imitations to Its Own Inventions." *Wall Street Journal,* December 1, 1981, p. 1.

Light, W. F. "The U.S. Market After the AT&T Consent Decree." *Signal,* Vol. 39, No. 5, January 1983, pp. 47–49.

"Linking Up to Boost the Office." *Business Week,* March 21, 1983, p. 144.

"L. M. Ericsson: Digesting the Microchip." *The Economist,* Vol. 277, January 16, 1982, pp. 68–69.

Loeb, G. H. "The Communications Act Policy toward Competition, A Failure to Communicate." *Duke Law Review,* No. 1, March 1978, p. 15.

Lomax, L. "Information, Please: Will Ma Bell Put Newspapers out of Business." *Texas Monthly,* Vol. 9, March 1981, pp. 98–101.

Lorenz, C. "AT&T Sets out to Take on the World." *Financial Times,* January 17, 1983, p. 8.

———. "Why ITT Finds that Being Big is Not Enough." *Financial Times,* June 6, 1982, p. 8.

Lowndes, J. C. "Optical Fiber Threatens Satellite Role in Voice Links." *Aviation Week and Space Technology,* Vol. 118, No. 1, January 3, 1983, pp. 62–65.

Lueck, T. "Amdahl is Still Guessing Right." *New York Times,* May 22, 1981, pp. D1/D4.

Lund, R. T. "Microprocessor and Productivity: Cashing in on our Chips." *Technology Review,* Vol. 86, No. 2, January 1981, pp. 33–54.

McAbee, M. "Pebbles Support Japan's Monolith." *Industry Week,* Vol. 197, No. 3, May 1, 1978, pp. 40–44.

McClellan, S. T. "See A Change In the Information Industry." *Datamation,* Vol. 28, June 1982, pp. 89–92.

McDonald. "Export Success Ranging from High Technology to Throat Lozenges." *Financial Times,* April 21, 1983, p. 12.

McKee, Edward A. "Local Distribution Networks: Safe from Competition." *Telephony,* Vol. 199, No. 9, April 20, 1981, pp. 22–24.

"The Mad, Mod, MUX World." *Datamation,* Vol. 27, December 1981, p. 67.

Magnet, M. "Clive Sinclair's Little Computer that Could." *Fortune,* Vol. 105, No. 5, March 8, 1982, pp. 78–83.

Manuel, T. "H.P.: A Dive Into Office Automation." *Electronics,* Vol. 54, November 3, 1981, pp. 106–210.

Manuso, D., and Barbetta, F. "Litton Awarded $276 M in Suit Against AT&T." *Electronic News,* Vol. 27, July 26, 1981, p. 1.

Marks, J. "ComSat, International Hotels to Offer International Teleconferencing." *Satellite Communications,* Vol. 6, No. 6, June 1982, p. 27.

Mason, B. "Pitney Bowes Into Word Processing." *Electronics,* Vol. 53, No. 2, January 17, 1980, p. 44.

Mayer, C. E. "Ex-Employee Urges Break-up of Bell System." *Washington Star,* May 27, 1981, p. F4.

Mayo, J. S. "The Power of Microelectronics." *Technology Review,* Vol. 86, No. 2, January 1981, pp. 46–50.

"MCI Files for Digital Termination Systems Network." *Data Communications,* Vol. 10, No. 12, December 1981.

Meadows, E. "Japan Runs Into America Inc." *Fortune,* Vol. 105, No. 6, March 22, 1982, pp. 56–61.

"Mid Continent Phone Seeks Cable TV Firms for Acquisition in 1980." *Wall Street Journal,* December 26, 1979, p. 14.

"Mitel: Telecommunications Success Story." *Financial World,* Vol. 150, November 15, 1981, p. 23.

"National Public Radio Talks up an Embryonic Digital Data Network." *Data Communications,* Vol. 11, No. 8, August 1982, p. 44.

"The New Sears." *Business Week,* November 16, 1981, pp. 140–41.

"New York Times Plans to Launch Satellite at California Plant." *Boston Globe,* April 7, 1982, p. 32.

Nichols, M. "Bell, Tex PUC Did Irreparable Harm." *Management Information Systems Week,* Vol. 113, June 10, 1971, p. 6.

"No. 1's Awesome Strategy." *Business Week,* June 8, 1981, pp. 86–88.

"NPR Set to Sell Satellite Time." *Broadcasting,* Vol. 102, January 11, 1982, pp. 75–76.

"Pacific Telephone's Product and Service Manager Approach to Competition." *Telephone Engineer and Management,* Vol. 77, No. 24, December 15, 1973, pp. 51–56.

Perlman, A. "Dataspeed 40/4—DP or Not DP—Appeal to Panel to Decide." *Electronic News,* Vol. 21, November 12, 1975, p. 30.

Perry, B. "Telco Equipment Depreciation Policies: Why Major Changes Are Needed—Fast!" *Telephony,* Vol. 199, No. 10, September 8, 1980, pp. 84–91.

Peters, T., and Waterman, R. "What's Right with Big Business?" *The Washington Monthly,* Vol. 14, No. 10, December 1982, pp. 37–45.

"Plessey Makes a Good Connection." *Financial Times,* September 27, 1982, p. 7.

Pollack, A. "Latest Technology May Spawn the Electronic Sweatshop." *New York Times,* September 3, 1982, p. 14.

Prescott, C. "Packaging a Genie." *Bell Telephone Magazine,* No. 5, 1982, p. 10.

Putka, G. "AT&T Depreciation Schedule Is a Major Issue." *Wall Street Journal,* January 13, 1983, p. 31.

"The Race to Plug In." *Business Week,* December 8, 1980, pp. 62–68.

"Regulation Rocks the Telex Market." *Business Week,* December 22, 1980, pp. 62–66.

Reston, J. "A Bonus for Congress." *New York Times,* December 15, 1982, p. A35.

Richter, M. J. "AT&T Rips ITT Quota Buy Proposal." *Electronic News,* Vol. 25, July 23, 1979, p. 1.

———. "Ex-Bell Chairman: We Thought Competition Would Lower Quality." *Electronic News,* Vol. 27, August 31, 1982, p. 21.

———. "FCC Rebuffs Fujitsu; OK's AT&T Fiber Pact to WE: Will Monitor Outside Buys." *Electronic News,* Vol. 28, April 3, 1982, p. 1.

———. "ITT Charges Bell WE Plan Violates Arm's Length Rule." *Electronic News,* Vol. 25, July 30, 1979, p. 21.

———. "Justice to Probe Cellular Arrangements." *Electronic News,* Vol. 28, July 12, 1982, p. 1.

"Rochester Telephone Expects Profit Gain in Six Months and Year." *Wall Street Journal,* March 17, 1982, p. 42.

Rose, F., and Chace, S. "IBM Agrees to Work with Mitel to Develop Phone Equipment for Entry in U.S. Market." *Wall Street Journal,* July 22, 1982, p. 4.

Rosenberg, R. "IBM Now Has a Say in the Store and Forward Message System." *Boston Globe,* September 24, 1981, p. 31.

————. "IBM Road Show Comes to Town." *Boston Globe*, April 24, 1983, pp. 77–78.

Ross, I. A. "New Technologies For the Exchange." *Bell Labs Record*, Vol. 60, April 1982, p. 83.

Saportio, B. "Kruger, The New King of Supermarketing." *Fortune*, Vol. 107, No. 4, February 21, 1983, pp. 76–80.

Saxton, W. A., and Edwards, M. "Bypassing Bell's Bottleneck." *Infosystems*, Vol. 28, No. 10, October 1981, p. 100.

————. "Data Role for Radio." *Infosystems*, Vol. 28, No. 12, December 1981, pp. 82–83.

Schuten, P. "Digital Dawn for Telephony." *New York Times*, December 31, 1978, p. D5.

"Sony: A Diversification Plan Tuned to the People Factor." *Business Week*, February 9, 1981, pp. 88–89.

"Standards Eased on Gear Added to Phone Network." *Wall Street Journal*, March 17, 1976, p. 20.

"Starting a Business—Pitfalls to Avoid." *U.S. News and World Report*, Vol. 91, No. 2, July 13, 1981, p. 75.

Stelzer, I. M. "The Post-Decree Telecommunications Industry." *National Economics Research Associates*, Princeton, N.J., May 11, 1982, pp. 1–19.

Stipp, D. "Hellooooo Electronics Federal Says." *Wall Street Journal*, April 20, 1982, p. 13.

"A Stronger Voice for Network Managers." *Business Week*, August 30, 1982, p. 62.

"Taking on the Industry Giant—An Interview with Gene M. Amdahl." *Harvard Business Review*, Vol. 58, No. 2, March/April 1980, pp. 82–93.

"Tandy Takes on the Telephone." *Business Week*, April 4, 1983, p. 68.

"Telecom Access is Where and When You Find It." *Telecom Times and Travels*, Vol. 2, No. 1, January 1983, p. 3.

Telecommunications Reports. Vol. 48, No. 10, March 8, 1982, p. 31; Statement of FCC Commissioner Anne P. Jones.

"Telecommunications Success Story." *Financial World*, Vol. 150, No. 21, November 15, 1980, pp. 23–24.

"Telepublishing in the Center Ring at ANPA: It's Everybody Else against AT&T." *Broadcasting*, Vol. 100, May 11, 1981, pp. 24–25.

"Teletype." *Fortune*, Vol. 5, No. 3, March 1932, pp. 40–43.

"Texas Court Hogties AT&T EIS Test." *Telephony*, Vol. 200, April 13, 1981, p. 16.

"The Thrill of Starting Up Again." *Business Week*, April 4, 1981, pp. 112–16.

"Tomorrow's Leaders: A Survey of Japanese Technology." *The Economist*, Vol. 283, June 19, 1982, pp. 5–32.

Tucker, W. "Public Radio Comes to Market." *Fortune*, Vol. 106, No. 8, October 18, 1982, pp. 205–10.

"Tymeshare Acquires Microband." *Wall Street Journal*, January 7, 1981, p. 33.

"Tymeshare Applies for Value Added Network." *Datamation*, Vol. 22, June 1976, p. 138.

Ungaro, C. "New and Old Players Set Stage for Networking Battle." *Data Communications*, Vol. 10, December 1981, pp. 86–87.

"U.S. Chief Named at Triumph-Adler." *New York Times*, February 2, 1982, p. D2.

"Users Turning to Packet Radio for Future Communications Needs." *Data Communications,* Vol. 9, August 1980, p. 34.

Uttal, B. "A Computer Gadfly's Triumph." *Fortune,* Vol. 105, No. 5, March 8, 1982, pp. 74–76.

———. "Texas Instruments Regroups." *Fortune,* Vol. 106, No. 3, August 9, 1982, pp. 40–45.

Vinton, R. "MCI Charges AT&T Tried to Put It Out of Business." *Electronic News,* Vol. 25, February 11, 1979, p. 1.

———. "WE Plant's Fate May Hinge on Plans for Merchant Market." *Electronic News,* Vol. 29, February 14, 1983, p. 50.

"Wall Street Journal Plans European Edition." *Editor and Publisher,* Vol. 115, May 1982, p. 62.

"Wang to Acquire Stake in SAT Firm." *Electronic News,* Vol. 28, November 22, 1982, p. 12.

"Wang to Enter CATV Industry." *Multichannel News,* Vol. 3, April 4, 1983, p. 1.

Wessler, B. "The Benefits of Private Packet Networks." *Telecommunications,* Vol. 14, No. 8, August 1, 1980, pp. 53–54.

"Where Future Jobs Will Be." *World Press Review,* Vol. 28, No. 3, March 1981, p. 23.

"A Whole New Way to Figure AT&T's Rates." *Business Week,* February 14, 1977, p. 86.

"Who Will Supply the Office of the Future?" *Business Week,* July 27, 1974, pp. 42–50.

"Why Ma Bell Won't Let a Western Contract Go East." *The Economist,* Vol. 281, November 7, 1981, p. 83.

Wiley, R., and Adams, D. "Should the FCC or USPS Control Electronic Mail?" *Legal Times of Washington,* Vol. 1, No. 46, April 6, 1979, p. 12.

BOOKS

AT&T. *Events in Telephone History.* New York: AT&T, 1971.

Bonsack, S. E. "A Discussion of Marketing in Telecommunications Under Conditions of Growing Competition." In *New Challenges to Public Utilities Management,* edited by H. Trebing. Institute of Public Utilities, Michigan State University. East Lansing: Michigan State University Press, 1974.

Brooks, John. *Telephone: The First Hundred Years.* New York: Harper and Row, 1975.

Casson, Herbert N. *The History of the Telephone.* Chicago: A. C. Mclury, 1910.

Conot, Robert. *A Streak of Luck.* New York: Simon and Schuster, 1978.

Coon, Horace, *American Tel. & Tel.* New York: Longmans, Green and Co., 1939.

Corfield, K. G. "Into the World of Broadcast Systems." In *Communications in the 1980s and After,* edited by J. Lighthill et al. London: The Royal Society, 1979.

De Sola Pool, Ithiel. "Retrospective Technological Assessment of the Telephone Report of the National Science Foundation." Vol. 1. *Research Program on Communications Policy,* Cambridge: MIT, 1977.

Howeth, L. S. *History of Communication—Electronics in the United States Navy.* Washington, D.C.: Bureau of Ships and Office of Naval History, 1962.

Irwin, Manley R. "The Communications Industry." In *Structure of American Industry,* 4th ed., edited by Walter Adams. New York: Macmillan, 1971.

Latske, Paul. *A Fight with an Octopus.* Chicago: Telephone Publishing Co., 1906.

Maclaurin, W. Rupert. *Invention and Innovation in the Radio Industry*. New York: Macmillan, 1949.

Macmeal, Harry B. *The Story of Independent Telephony*. Chicago: Independent Pioneer Telephone Association, 1939.

Mathison, S. and P. Walker, *Computers and Telecommunications: Issues in Public Policy*. Englewood Cliffs, N.J.: Prentice-Hall, 1970. Appendix C.

Osborne, Adam. *Running Wild: The Next Industrial Revolution*. Berkeley, Calif.: McGraw-Hill, 1979.

Paine, Albert Bigelow. *In One Man's Life*. New York: Harper, 1921.

Robinson, G. O., ed. *Communications for Tomorrow*. New York: Praeger Publishers, 1978.

Stehman, J. Warren. *The Financial History of the AT&T Company*. Boston: Houghton Mifflin, 1925.

Stelzer, I. *The Post-Decree Telecommunication Industry*. National Economic Research Associates, Inc., Princeton, N. J., May 11, 1982.

Todd, Kenneth P., Jr. *A Capsule History of the Bell System*. New York: AT&T, 1972.

Walsh, J. Leigh. *Connecticut Pioneers in Telephone*. New Haven, Conn.: Telephone Pioneers of America, 1950.

CONGRESSIONAL AND PRESIDENTIAL HEARINGS

Federal Communications Commission. *Investigation of the Telephone Industry in the U.S.* 76th Cong., 1st sess., House Document No. 340. Washington, D.C.: U.S. Government Printing Office, 1939.

National Commission on Electronic Funds Transfer. *EFT in the United States, The Final Report*. October 28, 1977. Washington, D.C., p. 166.

National Commission on Electronic Funds Transfer. *Suppliers Committee Public Hearings*, December 14, 1976. San Francisco.

President's Task Force on Communication Policy. *Final Report*. December 7, 1968.

Space Satellite Communications. *Hearings Before the Senate Subcommittee on Small Businesses*. 87th Cong., 1st sess., 1961.

Telephone Investigation, Pursuant to Public Resolution No. 8. 74th Cong. Washington, D.C.: U.S. Government Printing Office, 1938.

U.S. Congress, House. *Consent Decree Program of the Department of Justice*. Hearings Before the Antitrust Subcommittee (Subcommittee No. 5), 85th Cong., 2d sess., Part 2, 1958.

U.S. Congress, Senate. Industrial Reorganization Act. *Hearings before the Subcommittee on Antitrust and Monopoly*, Committee on the Judiciary. 93d Cong., 2d sess., Part 5. The Communication Industry, June 1974, pp. 2972–74.

REGULATORY DECISIONS, GOVERNMENT REPORTS

AT&T Revisions to Tariff, FCC Nos. 260 and 267, relating to Dataspeed 40.

CML Satellite Corporation. 51 FCC 2d 14, 1975.

FCC. Amendment of Section 64–702 of the Commission's Rules and Regulations (Second Computer Inquiry), Tentative Decision and Further Notice of Inquiry and Rule Making, Docket No. 20828. 1979, p. 70; *Final Decision, 1980*.

———. American Telephone and Telegraph Company. Regulations Relating to Connection of Telephone Company Facilities of Customers. Docket No. 12940, *Memorandum, Opinion and Order,* January 17, 1962.

————. American Telephone and Telegraph Company and the Bell System Operating Companies Tariff. FCC No. 8 (BSoc8), Exchange Network Facilities for Interstate Access (ENFIA). Docket No. 780371, *Memorandum, Opinion and Order*, April 1982.

————. Application of MCI, Incorporated for Construction Permits to Establish New Facilities in the Domestic Public Point to Point Microwave Radio Service in Chicago, Illinois, Saint Louis, Missouri, and Intermediate Points. Docket No. 10509, 1969.

————. Computer Inquiry. Tentative Decision and Further Notice of Inquiry and Rule-making, July 2, 1979.

————. Exhibit, Corporate Planning Report AT&T, 1972; Trial Staff Exhibit 147, 1974; also Trial Staff's Brief. Docket No. 19129 (Phase II), 2, 1976, p. 42.

————. The Future of Digital Technology in the Private Radio Services. Policy Division, Private Radio Bureau. Washington, D.C. August 1981.

————. In the Matter of Allocation of Frequencies in the Bands Above 890 Megacycles. Docket No. 11866, *Report and Order*, July 29, 1954.

————. In the Matter of Allocation of Frequencies in the Bands Above 890 Megacycles. Docket No. 11866, *Memorandum, Opinion and Order,* September 1960.

————.In the Matter of Amendment of Part 25 of the Commission's Rules and Regulations with Respect to the Procurement of Apparatus, Equipment, and Services Required for the Establishment and Operation of the Communication Satellite System and Satellite Terminal Stations. Docket No. 15123, *Report and Order,* April 3, 1964.

————. In the Matter of Amendment of Section 64.702 of the Commission's Rules and Regulations. Computer II Docket No. 20828, Reply Comments of AT&T, October 17, 1977.

————. In the Matter of American Telephone and Telegraph Company, the Associated Bell System Companies, Charges for Interstate Service, AT&T Transmittal Nos. 10939, 11027, 11657. Phase II, Trial Staff Testimony, M. R. Irwin, Docket No. 19129, 1975.

————. In the Matter of American Telephone and Telegraph Company, the Associated Bell System Companies, Charges for Interstate Telephone Service, AT&T Transmittal Nos. 10989, 11027, 11657. Docket No. 19129 (Phase II), *Final Decision and Order,* March 1977.

————. In the Matter of American Telephone and Telegraph Company and Associated Bell System Operating Companies for authority under Section 214 of the Communications Act of 1934 as amended, to supplement existing facilities by construction, acquisition and operation of a light guide cable between cities on a main route between New York, New York and Cambridge, Massachusetts and between Moseley, Virginia and Washington, D.C. File Ac. WDC 30 71. *Petition to Deny,* Fujitsu America, Inc., December 11, 1981.

————. In the Matter of Application of American Telephone and Telegraph Company, the Bell Telephone Company of Pennsylvania; the Chesapeake and Potomac Telephone Company, the Chesapeake and Potomac Telephone Company of Maryland, the Diamond State Telephone Company, New England Telephone and Telegraph Company, New Jersey Bell Telephone and Telegraph Company, New York Telephone Company, and the Southern New England Telephone Company. For authority under Section 214 of the Communications Act of 1934, as amended, to supplement existing facilities by construction, acquisition and operation of a light-guide cable between cities on a main route between Cambridge, Massachusetts and Washington, District of Columbia with extension lightguide cables to various cities along this route. F 16 No. W-P-C-3071. *Memorandum, Opinion and Order,* April 29, 1982.

———. In the Matter of the Application of Satellite Business Systems. File No. 7-DSS-D-76. Reply of ITT, September 24, 1976.

———. In the Matter of AT&T Petition for a Declaratory Policy That Advanced Communications Services May Be Provided Using Digital Facilities Heretofore Authorized by the Commission 1978.

———. In the Matter of AT&T's Revision to Tariff FCC Nos. 260 and 267. *Memorandum, Opinion and Order,* March 3, 1976.

———. In the Matter of Bell Operating Company Procurement of Telecommunications Equipment. Petition of ITT, May 30, 1979.

———. In the Matter of Bell System Procurement Practices, Comments of International Telephone and Telegraph Corporation, Docket No. 80–53, October 15, 1981.

———. In the Matter of Establishment of Domestic Communication-Satellite Facilities by Non-governmental Entities. Docket No. 16495, *Report and Order,* March 24, 1970.

———. In the Matter of Establishment of Policies and Procedures for Consideration of Application to Provide Specialized Carrier Services in the Domestic Public Point to Point Microwave Radio Service and Proposed Amendments to Parts 21, 43, and 61 of the Commission's Rules. Docket No. 18920, *First Report and Order,* June 3, 1971.

———. In the Matter of an Inquiry into the Administration and Regulatory Problems Relating to the Authorization of Commercially Operable Space Communications Systems. Docket No. 14024, *Second Report,* 1961.

———. In the Matter of MCI Telecommunications Corporation. Investigation into the Lawfulness of Tariff FCC No. 1 insofar as it purports to offer Execunet Service. Docket No. 2064a, *Decision,* July 13, 1976.

———. In the Matter of MTS and WATS Market Structure. Notice of Inquiry and Proposed Rulemaking. March, 1978.

———. In the Matter of MTS and WATS Market Structure, ICC Docket No. 78-72, Phase I, *Third Report and Order,* February 28, 1983, pp. 1–196.

———. In the Matter of New York Telephone Company Tariff No. 800. Dimension PBX—A Petition for Declaratory Ruling Under Relief. Computer and Business Manufacturers Association. June 16, 1978.

———. In the Matter of Presten Trucking Company, On Reconsideration of Grant Application for Microwave Facilities in the Motor Carrier Radio Service Inquiry into Certain Arrangements for Cooperative Use of Private Microwave. Docket No. 19309, *Memorandum, Opinion and Order,* August 24, 1971.

———. In the Matter of Regulatory and Policy Problems Presented by the Interdependence of Computer and Communication Services and Facilities. Docket No. 16979, *Final Decision and Order,* March, 1971.

———. In the Matter of Regulatory Policies Concerning Resale and Shared Use of Common Services and Facilities. Comments of Telenet Communications Corporation. Docket No. 20097. December 11, 1974, pp. 30–32.

———. In the Matter of Request for Declaratory Ruling and Investigation by Graphnet

Systems, Inc., concerning a proposed offering of Electronic Computer Mail ECOM. Docket No. 76–6, February 2, 1979.

———. In the Matter of Use of the Carterphone Device in Message Toll Telephone Service; In the Matter of Thomas F. Carter and Carter Electronics Corporation, Dallas, Texas. Complainants vs. American Telephone and Telegraph Company, Associated Bell System Companies, Southwestern Bell Telephone Company, and General Telephone Company of the Southwest. Dockets No. 16942 and 17073, *Final Decision*, 1968.

———. Letter to Dean Birch, Chairman, FCC, from Peter Flanagan, White House, January 30, 1970. Docket No. 16495. Appendix B, p. 23.

———. Regulatory Policies Concerning Resale and Shared Use of Common Carrier Services and Facilities. Docket No. 20097, *Report and Order*, 1976.

———. Reply Comments. OTP Executive Office of the President. Docket No. 297, February 24, 1975, pp. 17 and 18.

———. Report of the Telephone and Telegraph Committee of the FCC; In the Matter of the Domestic Telegraph Investigation. Docket No. 14650, April, 1966.

———. Special Investigation, Bell System's Patent Control: Its effect and suggested remedies. Docket No. 1, 1937, p. 10.

———. Testimony of M. R. Irwin, Trial Staff Exhibit. Docket No. 19129 (Phase II), 1975, p. 240.

U.S. General Accounting Office, Report. *Can the Federal Communications Commission Successfully Implement Its Computer II Decision?* January 29, 1982.

COURT DECISIONS

Computer and Communications Industry Association v. FCC et al. Civil Action No. 81-1193. U.S. Court of Appeals, District of Columbia. November 12, 1982.

CONRAC Corporation v. American Telephone and Telegraph Company, Telesciences Inc., Bell Telephone Company of Nevada et al. U.S. District Court, Southern District of New York. April 13, 1982.

GTE Service Corporation v. FCC. 474, F2d, 724, 1973.

Hush-a-phone Corporation v. American Telephone and Telegraph Company. 238 F 2d 266 (D.C. Cir - 1956).

IBM v. FCC. No. 77-4405. Second Circuit. January 4, 1977.

International Telephone and Telegraph Corporation v. American Telephone and Telegraph Company, Western Electric, Inc. and Bell Telephone Laboratories, Inc. U.S. District Court of New York. June 10, 1977. (Settled February 28, 1980.)

International Telephone and Telegraph Corporation v. General Telephone and Electronics Corporation and Hawaiian Telephone Company. Civil Action No. 2754. U.S. District Court, Hawaii. 1969.

International Telephone and Telegraph Corporation v. General Telephone and Electronics Corporation. 528FD. 913 (9th Circuit), 1975.

Litton Systems, Inc. et al. v. American Telephone and Telegraph, Inc. et al. Civil Action No. 1323-26, 1344. U.S. Court of Appeals, Second Circuit. February 3, 1983.

MCI Telecommunications Corporation et al. v. FCC et al. Motion for an Order Directing Compliance with Mandate. U.S. Court of Appeals, District of Columbia, April 1978.

MCI Telecommunications Corporation, Microwave Communications Incorporated and

N-Triple-6 Incorporated v. FCC et al. Petition for Review Orders of the FCC, U.S. Court of Appeals, District of Columbia, July 28, 1977.

North Carolina Utility Commission v. FCC. 537, F.dd 727, 1976.

Southern Pacific Communications Company et al. v. American Telephone and Telegraph Company et al. Civil Action No. 78-0545. U.S. District Court, District of Columbia. Memorandum Opinion of U.S. District Judge Charles R. Richey. December 21, 1982.

Telesciences, Inc. v. American Telephone and Telegraph Company, Bell Telephone of Nevada et al. U.S. District Court, District of Columbia. September 25, 1980. (Settled out of court.)

United States v. American Telephone and Telegraph Company et al. Civil Action No. 74-1690. U.S. District Court, District of Columbia. Brief of the People of the State of California and the Public Utilities Commission of the State of California on Issues Regarding the Absence of Restrictions on AT&T. June 13, 1982.

United States v. American Telephone and Telegraph Company, Western Electric Company, and Bell Telephone Laboratories, Plaintiffs, Third Statement of Contention and Proof. January 10, 1980.

United States v. American Telephone and Telegraph Company, Western Electric Company, Bell Telephone Laboratories. Civil Action No. 74–1698. U.S. District Court, District of Columbia. August 11, 1982.

United States v. Western Electric, Inc., AT&T. Civil Action No. 17-49. U.S. District Court, District of New Jersey. September 3, 1981.

Washington Utilities and Transportation Commission v. FCC. 1513 F.dd 1147, 1975.

PAPERS/CONFERENCES

Alleman, J., and E. Beauvais. "No Main Is an Island or Local Loops as Barriers to Entry." Paper delivered at the Western Economic Association Annual Meeting, July 5, 1981.

Crandall, R. W. "The Impossibility of Regulating Competition in Interstate Markets." Paper delivered at Eastern Economic Association, Boston, Massachusetts, May 12, 1979.

Hatfield, D. "Local Distribution—the Next Frontier." Ninth Annual Telecommunications Policy Research Conference, Annapolis, Maryland. April 29, 1981.

———. "Technological Transformation of Telecommunications and the Public Policy Response." Institute of Public Utilities Conference, Williamsburg, Virginia. December 3, 1980.

Hinchman, W. Remarks Before the ICA, Las Vegas, May 15, 1978.

Irwin, M. R. "Technology and Telecommunications: A Policy Perspective for the 80s." Working Paper No. 22, Regulation Reference. Ottawa, Canada, Economic Council of Canada.

———. "U.S. Telecommunications: Technology vs. Regulation." *Information Society: Changes, Chances, Challenges.* Fourteenth International TNO Conference, Rotterdam, the Netherlands, 1981, p. 47.

INDEX

About the Author

MANLEY RUTHERFORD IRWIN, a specialist in government regulation of business and industrial organization, is Professor of Economics in the Whittemore School of Business and Economics at the University of New Hampshire. He is the author of *The Telecommunications Industry*. His most recent contributions to books and journals include articles in *Telecommunications Policy* and *Public Utilities Fortnightly*, among others. He has served as a consultant to the U.S. and Canadian governments, testified before state and federal committees, and has lectured in Europe and the Far East.